LIVE LOOPING IN MUSICAL PERFORMANCE

Live Looping in Musical Performance offers a diverse range of interdisciplinary perspectives on the application of live looping technology by lusophone performers and composers. This book explores various aspects, including the aesthetic component, instrumentation, and setup, highlighting the versatility of this technology in music-making.

Written by musicians and researchers from Portuguese-speaking countries, this book comprises eleven chapters that delve into various musical contexts, genres, and practices. The novelty of including collaborative texts written alongside non-professional researchers offers the possibility of drawing from real experience to consider how live looping has been changing and "cyborguising" the concept of music, the ritual of the performance, the identity of the musicians, and the public's expectations.

Live Looping in Musical Performance provides cutting-edge reading for composers and performers, as well as ethnomusicologists, students, and researchers working in the areas of music production, technology, and performance. This book addresses a broader audience, both academic and non-academic, who are interested in new processes of musical creativity in a post-human world.

Alexsander Duarte is currently working as a Lecturer at the Federal University of Pará (UFPA) in Belém, Brazil. He is also a researcher at the "Music and Identity in the Amazon" Research Group at UFPA and a collaborator at the Institute of Ethnomusicology – Center for Studies in Music and Dance (INET-md), University of Aveiro, Portugal. As a musician and ethnomusicologist, his academic activities include concerts, as well as the publication of various forms of research output, such as CDs, ethnographic documentaries, articles, books, book chapters, and musical scores. His research interests encompass popular music in Lusophone territories, Intangible Cultural Heritage, and the application of audio technologies in sound production. As a postdoctoral researcher, he established LoopLab, a laboratory dedicated to live looping experimentation at the University of Aveiro.

Susana Sardo is Associate Professor of Ethnomusicology at the University of Aveiro in Portugal and Visiting Professor at Goa University in India, for the Cunha Rivara Chair. Her research interests include music and post-colonialism, sound archives, music in the Lusophone world, and the intersection of music and post-dictatorship regimes. Since 2013 she has been actively engaged in promoting shared research practices in ethnomusicology using research as a means for social transformation in the field. Susana Sardo is also the co-chair of the Study Group of Historical Sources for the International Council for Traditional Music (ICTM). In 2007, she founded the University of Aveiro branch of the Institute of Ethnomusicology – Research Center for Music and Dance (INET-md), which she coordinated until January 2023.

LIVE LOOPING IN MUSICAL PERFORMANCE

Lusophone Experiences in Dialogue

Edited by Alexsander Duarte and Susana Sardo

Routledge
Taylor & Francis Group
LONDON AND NEW YORK

Designed cover image: Álvaro Sousa

First published 2024
by Routledge
4 Park Square, Milton Park, Abingdon, Oxon OX14 4RN

and by Routledge
605 Third Avenue, New York, NY 10158

Routledge is an imprint of the Taylor & Francis Group, an informa business

© 2024 selection and editorial matter, Alexsander Duarte and Susana Sardo; individual chapters, the contributors

The right of Alexsander Duarte and Susana Sardo to be identified as the authors of the editorial material, and of the authors for their individual chapters, has been asserted in accordance with sections 77 and 78 of the Copyright, Designs and Patents Act 1988.

All rights reserved. No part of this book may be reprinted or reproduced or utilised in any form or by any electronic, mechanical, or other means, now known or hereafter invented, including photocopying and recording, or in any information storage or retrieval system, without permission in writing from the publishers.

Trademark notice: Product or corporate names may be trademarks or registered trademarks, and are used only for identification and explanation without intent to infringe.

British Library Cataloguing-in-Publication Data
A catalogue record for this book is available from the British Library

ISBN: 978-0-367-72259-3 (hbk)
ISBN: 978-0-367-72257-9 (pbk)
ISBN: 978-1-003-15408-2 (ebk)

DOI: 10.4324/9781003154082

Typeset in Sabon
by Apex CoVantage, LLC

CONTENTS

List of Contributors   *vii*
Acknowledgements   *xiii*

   Introduction   1
   Alexsander Duarte and Susana Sardo

PART I
Auto-ethnographic experiences (from the lab)   **11**

1   From the studio to the stage: Reflections on live looping
    and instrumentality through the performance of
    Import/Export: Percussion Suite for Global
    Junk by Gabriel Prokofiev   13
    Luís Bittencourt

2   HASGS: Its repertoire using live looping   34
    Henrique Portovedo

3   Interaction and reaction: Reflections about
    performance, composing, and live looping   51
    Iury Matias de Sousa

4   Viola Sertaneja and live looping in the performance
    of Antonio Madureira's Repente   66
    Erik Pronk

5 Quasitude: The processes and methods of the
  composition work for xylophone and live looping   78
  Helvio Mendes and Samuel Peruzzolo Vieira

6 Densus Bridge: For trumpet and live electronics
  (live looping and effects)   90
  Elielson da Silva Gomes and Alexsander Duarte

7 Perhaps the loop station is not the point   106
  José Valente

PART II
Collaborative writing experiences (from the field)   125

8 Creating atmospheres with LL: The forms
  of creativity of Tiago Oliveira   127
  *Melina Aparecida dos Santos Silva, José Cláudio
  Siqueira Castanheira and Tiago Oliveira*

9 The loop pedal and the guitar – individual practice
  and its connection to sociality   137
  *Ricardo Jorge Monteiro Cabral and Jorge Vicente
  dos Santos Almeida*

10 Taking the live out of looping – composing
   with the loop pedal   149
   Aoife Hiney and Isabel Novella

11 The one and the many: Interview with Portuguese
   singer and songwriter Joana Lisboa   163
   Samuel Peruzzolo Vieira and Joana Lisboa

Index   175

LIST OF CONTRIBUTORS

Luís Bittencourt is a percussionist, composer, artist-researcher, and music producer. Bittencourt holds a PhD and Master in Music Performance from University of Aveiro (Portugal), and a Bachelor Degree in Percussion Performance (Universidade Federal de Santa Maria – UFSM, Brazil). As a percussionist he has performed and collaborated with Lee Ranaldo (Sonic Youth), Jeffrey Ziegler (Kronos Quartet), Phill Niblock, David Cossin (Bang on a Can), Jon Rose, Emily Hall, among others. His pioneer academic investigations on the instrumentality of water and "found" instruments have established new insights on artistic experimentation in percussion music performance. Bittencourt is a member of Percussive Arts Society (PAS) and guest editor for journals Per Musi, SHIMA, and collaborates as revisor for the Society of Artistic Research (SAR) publications, among other.

Alexsander Duarte is currently working as a Lecturer at the Federal University of Pará (UFPA) in Belém, Brazil. He is also a researcher at the "Music and Identity in the Amazon" Research Group at UFPA and a collaborator at the Institute of Ethnomusicology – Center for Studies in Music and Dance (INET-md), University of Aveiro, Portugal. As a musician and ethnomusicologist, his academic activities include concerts, as well as the publication of various forms of research output, such as CDs, ethnographic documentaries, articles, books, book chapters, and musical scores. His research interests encompass popular music in Lusophone territories, Intangible Cultural Heritage, and the application of audio technologies in sound production. As a postdoctoral researcher, he established LoopLab, a laboratory dedicated to live looping experimentation at the University of Aveiro.

Aoife Hiney lectures at the Escola Superior de Educação (Instituto Politécnico do Porto), the University of Aveiro, Portugal and is an integrated researcher at Institute of Ethnomusicology – Center for Studies in Music and Dance (INET-md), University of Aveiro. Founding conducting of the choir Voz Nua, now a cultural association with four choirs under her direction, she also sings with the professional ensemble Záve and has been an associate director of the ZêzereArts Festival since 2013. From 2020 to 2021 she was the contracted researcher for the SOMA project at the University of Aveiro, promoting shared research practices in developing an archive of sounds and memories of Aveiro, financed by the Portuguese Foundation for Science and Technology. Aoife jointly coordinates Labeamus – Laboratory for the Teaching and Learning of Musics at the University of Aveiro. Her research focuses on choral singing, shared research practices, the Kodály Concept and non-professional musicians.

Joana Lisboa was born in Lisbon in 1983 and has always been passionate about music and singing. In 2007, at the age of 23, she performed for the first time as a guitar and voice duo in front of an audience. Since then, she has pursued various musical projects and developed her solo project, where she has honed her looping skills. To further her knowledge in music theory, and vocal technique, Joana has taken numerous courses on these subjects. She has also explored dance and theater. Since 2011, Joana has been a vocal coach, sharing her expertise in vocal technique with her students. In 2021, she completed her Community Music specialization at ESElx. In addition to her work as a voice and singing trainer, Joana is also a voiceover actress and a professional singer.

Iury Matias is a professional musician nominated for the 7th Profissionais da Música Award, 2023 edition, in four categories and has performed in several states of Brazil, Spain, Italy, Switzerland, Belgium, passing through Cape Verde and several cities in Portugal. Born in the city of Natal (Brazil), he graduated in music from the Federal University of Rio Grande do Norte. He is a music technician with a specialization in electric guitar (UFRN) and a master's student in musicology at the University of Aveiro. He was a music teacher, and runs a children's musicalization, guitar, and ensemble practice in regular private and public schools in Brazil. In Portugal, he taught guitar, harmony, ensemble class, and musical training for two years. He conceived and carried out the project of didactic concerts "Aula Show" in faculties, schools, social projects and cultural spaces in Brazil, Portugal and Cape Verde.

Helvio Mendes is a percussionist and researcher at Institute of Ethnomusicology – Center for Studies in Music and Dance (INET-md), University of Aveiro, and he is currently a Ph.D. candidate in Music at Aveiro University, Portugal. He earned his master's degree from the Universidade

Estadual de Campinas/UNICAMP, Brazil, and his bachelor's degree in percussion from the Universidade Estadual Paulista/UNESP, Brazil. Throughout his musical career, he has performed with several symphony orchestras and ensembles in the USA, Brazil, Canada, Portugal, Morocco, and Spain. In addition to being a founding member of Clube do Choro de Aveiro and part of the Strings+Bars Duo, he is also pioneering research on the xylophone. Helvio's research aims to find new artistic and expressive approaches for the instrument, which propose unusual sensorial experiences through sound experimentation of the xylophone with interactions using new technological tools.

Ricardo Cabral holds a degree in Anthropology and Sociology from Goldsmith's University in London. He is a Portuguese musician of Cape Verdean origin and works as a Learner Support Assistant at Access Creative College in London, UK. From 2013 to 2015, he was part of the research team for Skopeofonia, an ethnomusicology project of Institute of Ethnomusicology – Center for Studies in Music and Dance (INET-md), University of Aveiro, conducted in a Cape Verdean neighborhood in Lisbon that utilized shared research practices to promote social transformation through music research. He has over 15 years of experience as an independent musician and sound engineer, having recorded several artists and performed at numerous events in Portugal and abroad. His first solo album is named "Mal Famadu" and was released in 2017 with the promotion made by the international RDP Africa and the agency Music For All in Lisbon. Additionally, he has written various articles about social issues in Portuguese and Cape Verdean newspaper opinion columns.

Isabel Novella is a Pan-African singer, songwriter, and composer who has established a distinct style as a performing and recording artist. Her music is a fusion of African rhythms and Western influences, creating a harmonious balance between the two. Born in Mozambique, Isabel grew up performing on stages around the world, including celebrated international World Music and Jazz festivals. She has released two albums to date, actively participating as a composer and co-producer. In 2017, after many years of touring, she decided to pursue a degree in Music Production and is currently studying in Portugal. Isabel's creative process often involves solo work, where she records melodies using only her voice and constructs all elements of each song before adding any instruments. Her primary objective is to become a respected female singer, composer, and producer and to inspire women in Africa to take the lead in their own musical careers.

Tiago Oliveira is a versatile musician who has made significant contributions to the music industry in Angola. He is a multi-instrumentalist and plays a key role in the musical projects Ohali Music and Kosmik, where he has showcased his skills in playing various instruments such as guitar, bass, drums, and keyboards. In addition to his musical abilities, Tiago is also a music producer and entrepreneur. He is the owner of "Estúdio 2" in Luanda, Angola, a state-of-the-art music studio that has become a go-to destination for many artists and musicians in the country.

Samuel Peruzzolo Vieira is a Composer and music educator. Additionally, he is a researcher at Grupo de Pesquisa Arte Sonora/Mosaico (UFPA-Brazil) and at Institute of Ethnomusicology – Center for Studies in Music and Dance, having concluded in 2018 a Ph.D. in Musical Composition at the University of Aveiro-Portugal. Samuel is also a classically trained percussionist, having completed a M.M. from TAMU-C (USA), and a Bachelor at UFSM (Brazil). His area of activity is wide-ranging, revolving around musical composition and notation, aesthetics and performance studies. Samuel taught for over ten years as both composer and percussion teacher in universities, music conservatories, and other learning institutions in Portugal, USA, and Brazil. Winner of two international composition competitions, including the 2010 International Composition Contest of the PAS, his works are increasingly performed throughout America, Europe, and Asia. Samuel's compositions are published in the US by Musicon Publications and C. Alan Publications.

Henrique Portovedo is a professor at the University of Aveiro and a guest professor at the Real Conservatorio Superior de Música de Madrid. He holds a Ph.D. in the field of Science and Technology of the Arts (Performance and Computer Music) from the Portuguese Catholic University. Portovedo has also held various research positions, including as Fulbright Researcher at the University of Santa Barbara California, an Erasmus Researcher at the University of Edinburgh, and a visiting researcher at both the ZKM Karlsruhe and McGill University Montreal. Portovedo has been awarded several prizes, including from the Portuguese National Centre of Culture and the British Society for Education Music and Psychology. As a performer and intermedia artist, Portovedo has presented multidisciplinary creations in festivals worldwide, and he has performed as a soloist with some of the most important contemporary ensembles in Europe. He is the coordinator of the research group "Creation, Performance and Artistic Research" of Institute of Ethnomusicology – Center for Studies in Music and Dance (INET-md).

Erik Pronk is a guitar performer and professor at the Federal University of Paraíba (UFPB). He earned a master's degree in classical guitar performance from the Royal Conservatoire of The Hague in the Netherlands and a Ph.D. in music from the University of Aveiro in Portugal. His academic research has primarily focused on solo guitar repertoire, but he has also worked with chamber music and arranging for various ensembles. At UFPB, Pronk has worked with the Iamaká group, and the Camerata Filipeia, a group of plucked chordophone players who focus on Armorial and Andean music. Pronk's doctoral research focused on the performance of Armorial music repertoire using loops, a movement that originated in northeastern Brazil in the 1970s and featured the viola as an important instrument due to its connection with Brazilian popular culture.

Jorge Almeida was born in São Vicente, Cape Verde. He is a guitarist, musical arranger, producer, and performer who works independently in Portugal and is affiliated with music labels end promoters Klassik and Cavi. With a passion for music and an extensive background in the industry, Jorge has developed a reputation for his exceptional talent and creativity in playing the guitar. Jorge received his education at Rojero Lima School in São Vicente, Cape Verde. He first worked as a session musician in various hotels on Boa Vista's island, Cape Verde, where he developed his music theory and performance skills. He later honed his skills through years of dedicated practice and performance, allowing him to create a unique style that blends traditional Cape Verdean music with contemporary influences. He has been on stage with influential artists such as Eneida Marta, Mayra Andrade, Dino D'Santiago and Ana Moura.

Melina Santos received her Ph.D. degree in Communication from Federal Fluminense University – UFF (Brazil) in 2018 and is currently working as a Postdoctoral Researcher at the Graduate Program in Culture and Territorialities at the same institution. In addition, she has also worked as a Postdoctoral Researcher at the Graduate Program in Communication at Pontifical Catholic University of Rio Grande do Sul – PUCRS in 2022. Her research interests include the Culture of Music Genres, African Metal Music Scenes, Africanisms in Music Production, Intersectionality and Media Production, Decoloniality, and Post-Colonial Theory. Furthermore, she is the author of the book entitled "We do rock too: Formas de criatividade do rock angolano" (EDUERJ, 2022).

Susana Sardo is Associate Professor of Ethnomusicology at the University of Aveiro in Portugal and Visiting Professor at Goa University in India, for the Cunha Rivara Chair. Her research interests include music and postcolonialism, sound archives, music in the Lusophone world, and the intersection of music and post-dictatorship regimes. Since 2013, she has been actively engaged in promoting shared research practices in ethnomusicology using research as a means for social transformation in the field. Susana Sardo is the co-chair of the Study Group of Historical Sources for the International Council for Traditional Music. In 2007, she founded the University of Aveiro branch of the Institute of Ethnomusicology – Center for Studies in Music and Dance (INET-md), which she coordinated until January 2023.

Elielson da Silva Gomes is currently a Ph.D. student and a member of the Loop Laboratory at the Institute of Ethnomusicology – Center for Studies in

Music and Dance (INET-md), University of Aveiro. Additionally, he is a professor of trumpet at the School of Music at the Federal University of Pará. He graduated in composition from the Federal University of Pará, holds a bachelor's degree in trumpet from the State University of Pará, and completed a technical course in trumpet, production, and cultural design from the Carlos Gomes Foundation. He served as principal trumpet for the Theatro da Paz Symphony Orchestra and the Amazonia jazz band, both located in Belém do Pará. He represented Brazil in the 2000 Olympic Band at the Sydney Olympic Games in Australia, where he received three awards. In 2014, he participated as a musician and arranger in the Brazilian Trumpet Ensemble and was a guest at the 39th Annual Conference of the International Trumpet Guild in the USA.

José Cláudio Castanheira holds a Ph.D. in Communication from Fluminense Federal University in Brazil, where he also serves as a professor in the Communication Department. Additionally, he is a professor at the Postgraduate Program in Communication in the Federal University of Ceará (UFC). During his doctoral studies, he completed a research internship at McGill University in Canada. In 2021–2022, he conducted research at the Communication Department of Universitat Pompeu Fabra in Barcelona. Castanheira is the leader of the research group GEIST (Study Group on Images, Sonorities and Technologies) formed by researchers from six Brazilian institutions: UFSC, UFES, IFRS, UFRGS, UNISINOS, and UFF, and he specializes in digital culture, music, sound studies, and film studies.

José Valente is a violist, improviser and composer. He holds a Ph.D. in contemporary Art by the Royal College for the Arts, University of Coimbra. Considered as one of the most innovative violists because of his eclectic approach, he has played his music in several stages and festivals worldwide, from India to the U.S.A. He has performed with musicians such as Paquito d'Rivera, Dave Douglas, Jason Kao Hwang, Shakir Khan, Melanie Charles or Alberto Conde. Among his many compositions, he created "Passaporte" for RTP/Antena 2, "Trabalho, Som, Palavra, Preguiça" for MAAT, in collaboration with writer Gonçalo M. Tavares. He has also released several albums, including the first album ever for viola and marching band ("Trégua"). In recognition of his work, Valente has been awarded with several prizes, including the Carlos Paredes Award for his album "Serpente Infinita", or The Hannah S. and Samuel A. Cohn Memorial Foundation Endowed Fellowship (USA).

ACKNOWLEDGEMENTS

English review: Briony Andrews

INTRODUCTION

Alexsander Duarte and Susana Sardo

Why live looping?

The concept of "loop" has been adopted and discussed in different areas of knowledge such as philosophy, visual arts, mathematics, economics, and sound studies. In the field of music, the feature designated as a loop – short and repeated segments of audio, that can be overlapped – especially in musical compositions and performances, dates back to experimental works developed with magnetic tape in the 1950s and 1960s, namely the minimalist works of Terry Riley and Steve Reich. However, the term became more common after the development of audio technology that allows musicians to electronically add pre-recorded repetitive layers of short melodic or rhythmic cells to their compositions in live performances.

In the 1970s, Brian Eno and Robert Fripp teamed up to produce the album "No Pussyfooting" (1973) to which they used the "Time-Lag Accumulator" system created by Terry Riley, by manipulating two tape recorders. Robert Fripp decided to rename this system initially as "frippertronics" and later as "soundscapes", thus adopting Michael Southworth's proposal applied to the "Listening to the Cities" (1969). This concept was later elaborated by the composer and music theorist Raymond Murray Schafer (1977), who transformed it into one of the most important theoretical concepts in the field of sound studies (Sterne 2012). In the following decades, the technique of using pre-recorded loops played simultaneously with acoustic sounds produced in real time was incorporated both by composers of so-called "electronic music", "electroacoustic music" or "mixed music", as well as by composers of popular music such as rock, jazz, hip-hop, etc. Nowadays, the term live looping commonly refers to a technology used in musical composition and performance that utilises loops recorded in real time. This technology – which can be explored from hardware, or software installed on laptops or devices such as tablets or smartphones – spread from the early 2000s onwards with the advent of the loop pedal.

In music, the use of the loop technique in general and live looping in particular has grown considerably over the years, promoting the emergence of a community of practice (Wenger 1998) composed of performers that refer to themselves as *loopers*, *loop artists*, or *one-man bands* (in the case of individual performers who use multiple instruments). Examples of some internationally known loopers are Dub FX, Reggie Watts, Ed Sheeran, Matthew Schoening, Josie Charlwood, Kewahi, Gavin Castleton, Andrew Bird, Yellow Ostrich, Grace McLean, Noiserv, and Thiago Ramalho, among others. The considerable number of loopers with musical works and other resources published on the web can be verified on two main electronic platforms: Looper's Delight and Livelooping.org.[1] The looper community of practice has also promoted several iconic festivals such as the Y2K International Live Looping Festival, the Loop Jubilee and the LiveLooping Festival in Cologne, Germany. The growth of this movement has significant repercussions on the sound technology industry, and some commercial companies not only sponsor the most renowned loopers, but also promote and support competitive events around the world as a way of publicising their products.

Research on loop technology and its impact on ways of making music and organising performative human worlds is relatively recent. In fact, most academic studies interested in the live looping phenomenon emanate from sound studies or popular music. The first deals with research that mainly addresses sound technologies, while popular music studies focus on sociological approaches associated with the music industry. In the field of ethnomusicology, this book is probably one of the first publications that focuses on live looping. It is the result of an experience developed at the University of Aveiro (Portugal) between 2016 and 2021, financed by the Portuguese Foundation for Science and Technology and the Institute of Ethnomusicology (INET-md[2]), which led to the creation of a Laboratory of Live Looping, the LoopLab.[3] The purpose of LoopLab is to carry out ethnographic research on different artistic projects by artists from Portuguese-speaking countries who use live looping technology as their main resource. At the same time, it aims to offer the possibility of artistic experimentation developed in a laboratory environment, using shared research practices (Sardo 2017) between ethnomusicologists, artists, and post-graduate students.

After five years of activity, LoopLab is now a very important host for Master's and PhD students, and a trigger for organising concerts and academic events like the first LiveLoopists Meeting held at the University of Aveiro in 2018, with this book being one of its outputs. Most publications that resulted from the work carried out in LoopLab are focused on updating the state of the art in the field (Duarte 2020), but are also related to particular experiments carried out by researchers, such as the application of live looping and electronics to xylophone performance (Mendes, Duarte, & Traldi 2019); implications of live looping and electronics in real-time performance (ibidem 2018a, 2018b); or the way the loop pedal can be defined as a new musical instrument (Duarte, 2016).

More ground-breaking research is the PhD thesis of Erik Pronk intitled "From romance to the north-eastern loop: viola *sertaneja* and live looping in armorial music" (Pronk, 2021). Pronk, who also authored a chapter in this book, shows how the experiences of Pierre Schaeffer, Pierre Henry, Terry Riley, Steve Reich, Brian Eno, and Robert Fripp contributed to the results of his research, in particular the use of the "time-lag-accumulator" system and the delay effect.

This book, together with the publications above, aims to advance the state of the art and include an ethnomusicological perspective based on the ethnography of live looping practices, articulated with artistic and technological laboratory experiences and collaborative research. One of our main contributions is to rethink the concept of performance in music proposed by Thomas Turino. In his book *Music as Social Life* (2008), Turino suggests approaching musical performance in the West from two broad perspectives: on the one hand, "live performance" (participatory and presentation), and on the other hand, "recording performance" (high fidelity and studio audio art). However, live looping technology has demonstrated that studio music production techniques have been incorporated into real-time performance, which promotes a new paradigm of musical performance that, in turn, recursively influences technical and aesthetic conceptions of the studio's music production (Knowles & Hewitt 2012). Knowles and Hewitt define this new musical paradigm as "performance recordivity". This type of performance allows the audience to systematically perceive the construction of sound blocks, as a denudation of the laboratory process of "studio audio art". Thus, there is an emphasis not only on the result, but also on the process that led to the performative outcome.

But this book also intends to show how research on music can contribute to the interdisciplinary debate related to the "cyborgization" of the world through artificial intelligence. Our research on loop and loopers pointed out the fact that industry also designates a machine that can trigger the loop (the loop pedal) as a looper. The coincidence that both the musician and the machine are given the same name cannot be underestimated. The transformation of human work into the work of the machine has been widely discussed, but the mixing, fusion, or blending of humans themselves with machines is one of the most challenging debates in social, ideological, ethical, and artistic fields. The fact that, in the case of live looping in music, musician and machine are termed the same way (loopers) adds an important novel argument to the debate. As Donna Haraway points out in her premonitory "Cyborg Manifesto", in advocating for an ideological world not based on gender, the combined human/machine instance of the cyborg is an important signal for discarding the dichotomy between nature and culture. According to Haraway:

> Late twentieth-century machines have made thoroughly ambiguous the difference between natural and artificial, mind and body, self-developing and externally designed, and many other distinctions that used to apply to organisms and machines.
>
> *(Haraway 1991: 152)*

In the dichotomic list used by Haraway to discuss the transformation of the world from a "comfortable hierarchical domination" system into the "informatics of domination", the author also includes the transformation of the mind in artificial intelligence as an eminent paradigm of conviviality:

> A cyborg world might be about lived social and bodily realities in which people are not afraid of their joint kinship with animals and machines, not afraid of permanently partial identities and contradictory standpoints. The political struggle is to see from both perspectives at once because each reveals both dominations and possibilities unimaginable from the other vantage point.
>
> *(Haraway 1991: 154)*

"Live Looping in Musical Performance: Lusophone Experiences in Dialogue" offers an important contribution for the above issues by including contemporary research on the use of looping technology in musical performance within the framework of ethnomusicology and artistic research. The book contains eleven chapters written by performers and/or composers, in which they present their modus operandi for exploring looping technology, both as a compositional tool and as a live performance technique, focusing on different aspects, from the aesthetic analyses to instrumentation and even setup. Voice, saxophone, viola *caipira* (Brazilian ten-string guitar), guitar, trumpet, percussion, and unconventional instruments are combined with pedals or other peripheral devices, to create a sound result that ranges from "traditional music" to "experimental" music. The composers explore different procedures throughout the compositional processes and in the adopted methodologies to achieve the desired sound and performance results. While some of them organised their composition projects with the aim of using the loop pedal, others chose to use software such as Pure Data, MAX/MSP, or others, developed exclusively for this purpose. In some chapters, the authors also address the technical issues associated with sound processing, such as the audio signal flow, and technical resolutions from the gesture point of view in interpretation and musical performance.

The book is divided into two parts. The first one, entitled Auto-ethnographic experiences (from the lab), consists of seven chapters written by performers who are also professional researchers. The second part, entitled Collaborative writing experiences (from the field), contains four chapters written collaboratively between musicians and professional researchers who, in most cases, are also musicians. This option by the editors seeks to focus on knowledge production processes organised through the use of "shared research practices" (Sardo 2017; Miguel, Rannochiari, and Sardo 2021), a methodological tool that also includes collaborative writing, and whose contribution is focused not only on the research product but also on the process. Shared research practices promote mutual learning between the different subjects involved in the research (professional researchers and non-professional researchers), which offers them equal access to knowledge of the worlds that each one represents. In this sense, multi-sited fieldwork was developed and intermediated by researchers who are members of the INET-md and specialists in the "musical scenes" (Cambria 2017) of the Portuguese-speaking territories covered in this work, comprising Portugal, Brazil, Cape Verde, Angola, and Mozambique.

Auto-ethnographic experiences (from the lab)

In **Chapter 1**, percussionist and researcher Luís Bittencourt proposes a deeper epistemological reflection on looping technology as a tool for creating and performing music. His contribution addresses crucial issues such as the discussion on the concept of instrumentality applied to the loop pedal, how live looping changes our perception of musical performance, and how looping technology dilutes the difference between studio and stage by transforming the stage into a kind of studio. Bittencourt's arguments are supported by artistic research that he puts into dialogue with the percussionist Joby Burgess in a very refined way.

In **Chapter 2**, composer-performer and researcher Henrique Portovedo discusses live looping as a compositional tool in electronic and mixed music. Centred on his own research while creating the HASGS project – Hybrid Augmented Saxophone of Gestural Symbiosis – the author compares, through a very accurate description of experimental work with composers, the strengths and weaknesses of looping technology with regard to working with electronic instruments or with acoustic instruments, aiming to integrate live looping to produce new sound and aesthetic realities in contemporary music. In the case of Portovedo, the researcher is also the composer and the performer, which increases the critical potential of addressing live looping tools.

Although live looping has been addressed by most of the authors as an excellent tool for new creative actions, it can also represent problems and difficulties. This is what Iury Matias refers to in his text about his experience as a songwriter. In **Chapter 3** Matias dedicates part of the text to describing his experience as a composer while trying to aggregate previous collective projects into a single authorial work and performance. This process completely changes the composition design, as now the creative ideas arise from the relationship with the looping technology, and not only from using live looping to develop previous creative ideas. However, the complexity of the looping technology and the huge possibilities it offers, limit the capacity of the composer and performer to dominate all the tools. In addition, the portability of the loop pedal decreases with its complexity, and it requires conditions for live performance which are not solely dependent on the performer or the composer. Matias offers in his chapter an excellent portrayal of the life of a composer when looping technology becomes a principal tool for composition.

Erik Pronk, researcher and instrumentalist, proposes a new approach to live looping by including it as an important musical arrangement tool. In a very innovative process, the author of **Chapter 4** connects the looping pedal with a traditional Brazilian instrument (viola *sertaneja* or *caipira*) for the interpretation of repertoire integrated in the Armorial Movement of the 1970s, which claimed the creation of a Brazilian musical style based on the African and Iberian heritage of local north-eastern Brazilian music. Adding the use of a contemporary electronic device to perform and recreate Armorial music is a huge challenge that is very well substantiated by the author through an interesting dialogue between ethnomusicology and artistic research methods.

Chapter 5 is an inspired proposal for collaborative work between instrumentalist and composer. Hélvio Mendes, Brazilian xylophonist, and Samuel Peruzzolo, Brazilian composer, both researchers at LoopLab, undertook synergistic work in order to create together a composition and performance for which looping technology was an essential tool. The loop pedal, in this case, had a great contribution not only in expanding the possibilities of the acoustic instrument, but also to breaking down hierarchies between composer and performer, thus contributing to more enriching compositional constructions.

Elielson Gomes, researcher and trumpeter, and Alexsander Duarte, ethnomusicologist and multi-instrumentalist, carried out a truly collaborative project based on experimental processes of applying the loop pedal to trumpet performance. Their experience is very well described in **Chapter 6** and the great contribution of this chapter focuses on the use of live looping by a musician who has never had contact with electronic media. Connected mainly to classical music and popular music, Elielson was introduced to looping technology by Duarte, and both created a work that is also a kind of matrix adaptable for instruments other than the trumpet. In this case, looping technology transformed not only the work and the instrument, but also the musician.

Chapter 7 is a journey through the creative process of its author, José Valente. Valente describes in detail how he uses looping technology in his compositions and subsequent performances as an intimate tool closely linked to his creative persona. This chapter is a kind of autoethnographic exercise that addresses concerns, risks, unquietness, ambiguities, and the paths taken by the composer–performer to reach an "emancipatory level of consciousness" regarding his own daily life.

Collaborative writing experiences (from the field)

Chapter 8 opens the second part of the book dedicated to collaborative texts written between academic and non-academic musicians. Melina Santos, a researcher in the field of ethnomusicology, communication and the music industry, and José Cláudio Castanheiro, a researcher in the field of communication, image studies, sonorities and technologies, write together with Tiago Oliveira, a producer and multi-instrumentalist from Angola who uses the loop as a way to create "sound atmospheres". The text explores how the live looping tool offers creative solutions for musical expression for instrumentalists and singers who consider themselves to have low musical literacy. However, as Santos, Castanheira and Oliveira argue, live looping opens up incredible creative avenues where the "musical gesture" is at the centre of the discussion.

In **Chapter 9**, Ricardo Cabral, sociologist and musician, proposes an intimate journey through the dialogue between live looping and the guitar of Jorge Almeida (Djode), a Cape Verdean musician in the diaspora. Moreover, this text proposes a deep reflection on live looping in music as a political expression, as a decolonial tool, and as a possibility for a musician from the global south to situate himself in the global north, where he can also impress his own African identity.

While the first two collaborative texts used an initial interview as the trigger for writing the text, in **Chapter 10**, Aoife Hiney – researcher in the field of community music and conductor – and Isabel Novella – singer and songwriter of Mozambican origin – decided to write the text in a very collaborative way. In this chapter, with a strong ethnographic profile, both authors describe deeply how live looping offers the possibility for a singer-songwriter to overcome an inability to read and write music. Live looping opens a pathway for "freedom", a possibility for Isabel, as a composer, to express her "musical ideas" to other musicians when she needs to transform her creative ideas into a real performance.

But the loop can also be understood to be a tool for extending the voice, and modifying and multiplying the musical personalities of the composer and performer. This is what Samuel Peruzzolo – researcher and composer from Brazil – and Joana Lisboa – composer and singer from Portugal – explain in **Chapter 11**. Live looping is a way of simultaneously hiding and exposing subjectivities, an opportunity to model the body, and also a tool for transformation and social intervention. The analysis of a project and three songs written by Lisboa explains very well how the musician uses live looping as a concept to shape her own life.

The analysis of the texts presented in this book leads us to interesting conclusions. The first, and probably most obvious, is that looping technology is not just a tool, but a non-human device that is transformative for the contemporary music scene. It potentially transforms key concepts like music, performance, composition, musician, and audience. The act of creating music becomes inspired and, to a certain extent, controlled by what looping technology can offer musicians, sometimes in unexpected ways. In the case of the dialogue between musicians and live looping, "creation is much more widely distributed" (Hennion 2003: 6) because looping technology works as a true mediator, that is, an instance with the agency to transform (Latour 2005). Loop tools, of course, do not make music, people do. Loop tools are electronic devices that result from layers of mediations of multidisciplinary origin. All those layers have the power to "transform, translate, distort and modify the meaning of elements they supposedly carry" (Latour 2005: 39). What loop technology carries are electronic circuits that can "potentially" produce sound and music.

So, although the loop pedal is not a musical instrument, it can also work as a musical instrument. Simultaneously, it has the power to expand the performative possibilities of conventional musical instruments and of the musician's body, but it can't have an autonomous life in the sense of producing sound and music only from its electronic circuits. Human action is always necessary, be it in the production of the technology or in the construction of the pre-recorded sounds or, even, in the way that they can be processed or combined. Loop tools can be an extension of the instrumentalist, and a possibility for those who have creative musical ideas but who cannot use a conventional instrument due to a lack of musical literacy. In a way, live looping has the power to turn amateurs into musicians and even professional musicians. And this is a big change in the way we understand the profile of the musician as a product of a long disciplinary maturation. Live looping, in this case, democratises music because it is already accessible to everyone, especially in the Anthropocene era which is also the era of acceleration, of the digital revolution where novelty is not an aspiration but a mandatory desideratum. As far as creativity is concerned, live looping has infinite options, and the more the musician knows the technology, the more they take advantage of its sound and musical possibilities, increasing their own creativity.

For both musicians and audiences, the perception of musical performance changes radically when the loop tools are on stage. Live looping blurs the difference between studio and stage by turning the stage into a studio and presenting the interstices between performance and composition to the audience's eyes and ears in real time. Therefore, it also democratises the performance and promotes more accurate listening due to the audience's attention to and curiosity about the unexpected result of using the looping technologies. The electronic profile of loop tools and, in a way, their cyborg identity, also raises the expectations of the various individuals who participate, especially in live performances, of their performative character. In this sense, live looping practices can also be a way of creating new subjectivities, and a political tool for liberation and social transformation by unveiling an important process where humans and machines unexpectedly unite and merge to make music. The disturbing thing is that it is this blend of human and machine – both called loopers – that reduces inequalities between composer and performer, makes new musical bodies emerge, offers new possibilities for the public to understand music and, for musicians, creates the illusion of not being alone.

Acknowledgement

Alexsander Duarte would like to acknowledge his FCT post-doctoral grant (SFRH/BPD/111870/2015).

Notes

1 Both electronic platforms can be seen in the following links: Looper's Delight (https://loopers-delight.com/) and LiveLooping.org (http://www.livelooping.org/).
2 http://www.inetmd.pt/index.php/en/.
3 https://www.facebook.com/LoopLabUA/.

References

Cambria, Vicenzo. "Cenas musicais": reflexões a partir da etnomusicologia". *Música e Cultura – Revista da Associação Brasileira de Etnomusicologia* 10 (1) (2017): 77–93. Retrieved from: https://www.abet.mus.br/volume-10-2017/.

Duarte, Alexsander. "A arte do looping: a loop station como instrumento de prática performativa musical. *Revista Post-ip* 3 (2016): 9–19. DOI: https://doi.org/10.34624/postip.v3i3.1585.

Duarte, Alexsander. "O uso da Loop Station em performance musical: implicações e exigências interpretativas". *Vortex Music Journal* 8 (2020): 1–22. DOI: https://doi.org/10.33871/23179937.2020.8.2.14.

Haraway, Donna. "A Cyborg Manifesto: Science, Technology, and Socialist-Feminism in the Late Twentieth Century," in *Simians, Cyborgs and Women: The Reinvention of Nature* (New York: Routledge, 1991), 149–181. (First published in *Socialist Review* LXXX, 1985.)

Hennion, Antoine. "Music and Mediation: Towards a new Sociology of Music". In *The Cultural Study of Music: A Critical Introduction*, M. Clayton, T. Herbert, R. Middleton (London: Routledge, 2003), 80–91.

Knowles, Julian & Hewitt, Donna. "Performance recordivity: studio music in a live context". In *7th Art of Record Production Conference*, Richard James & Katia Isakoff (2012). Retrieved from: https://eprints.qut.edu.au/48489/.

Latour, Bruno. *Reassembling the Social. An Introduction to Actor-Network Theory* (Oxford: Oxford University Press, 2005).

Mendes, Hélvio, Duarte, Alexsander, & Traldi, César. "XyLoops: Composition and performance of a work for xylophone and live electronics (live looping)". *Research Highlights, Revista Research@UA*, 9 (2019): 59–60. Retrieved from: https://issuu.com/gai_ua/docs/research_2018.

Mendes, Hélvio, Duarte, Alexsander, & Traldi, César. "Xyloops: Composição e performance de uma obra para xilofone e eletrônica em tempo real (live looping)". *Vortex Music Journal*, 6 (2018): 1–20. Retrieved from: https://periodicos.unespar.edu.br/index.php/vortex/article/view/2609.

Mendes, Helvio, Duarte, Alexsander, & Traldi, César. "(Original Score): XyLoops to Xylophon and live electronics (Live Looping)". *Vortex Music Journal*, 6 (2018). Retrieved from: https://periodicos.unespar.edu.br/index.php/vortex/article/view/2623.

Miguel, Ana Flávia, Ranocchiari, Dario, & Sardo, Susana. "Shared research practices in music. Attempts and challenges from Portugal." *Revista de Antropología Iberoamericana* 15, no. 2 (2020): 357–382. DOI: 10.11156/aibr.150208e. Retrieved from: https://www.aibr.org/antropologia/netesp/numeros/1502/150208e.pdf.

Pronk, Erik de Lucena. *Do romance ao loop nordestino: viola sertaneja e live looping na música armorial*. (Doctoral Dissertation, University of Aveiro, Portugal, 2021). Retrieved from: https://ria.ua.pt/handle/10773/33121.

Sardo, Susana. "Shared Research Practices on and about music: toward decolonizing colonial ethnomusicology". In *Making Music, Making Society*, edited by Josep Martí and Sara Revilla, 217–238. Cambridge: Cambridge Scholars, 2017.

Schafer, Murray. *The Tuning of the World*. 1st ed. (New York: A.A. Knopf, 1977).

Southworth, Michael. "The Sonic Environment of Cities". *Environment and Behavior* 1, no. 1 (1969): 49–70.

Sterne, Jonathan. *The Soundstudies Reader* (London: Routledge, 2012).

Turino, Thomas. *Music as Social Life: The politics of participation* (Chicago: The University of Chicago Press, 2008).

Wenger, Etienne. *Communities of practice: Learning, meaning, and identity* (Cambridge: Cambridge University Press, 1998). DOI: https://doi.org/10.1017/CBO9780511803932.

PART I
Auto-ethnographic experiences (from the lab)

PART I

Auto-ethnographic experiences (from the lab)

1
FROM THE STUDIO TO THE STAGE

Reflections on live looping and instrumentality through the performance of *Import/Export: Percussion Suite for Global Junk* by Gabriel Prokofiev

Luís Bittencourt

Introduction

In our current digitally-driven era we find a variety of devices, mediums, systems, and interfaces that are challenging past concepts of what a musical instrument may be. It is common to think of musical instruments as "discrete, self-subsisting material objects, intentionally crafted for the purpose of making music by performing musicians" (Alperson 2008, 38), or even through more comprehensive concepts that classify musical instruments as "sound producing devices" (Hornbostel 1933: 129; Lysloff & Matson 1985: 217).

Digital music instruments such live looping devices, on the one hand, are musical artefacts intentionally crafted to make music; on the other hand, they do not produce sounds, being recognised more as sound reproducing devices. Additionally, the performance with live loopers requires the mastering of their inherent performance techniques (just as any musical instrument does), but the type of virtuosity associated with these devices may differ from that normally identified in the performance of standard musical instruments.

This chapter reflects on the instrumental potential of looper devices through a discussion on the concept of instrumentality and a case study on the performance of *Import/Export: Percussion Suite for Global Junk* by Gabriel Prokofiev, a work composed for a quartet of junk objects, live looping and sound processing. Through a performative research (Haseman 2006), led by my own artistic practice in the performance of *Import/Export*, plus interviews with the composer and percussionist Joby Burgess,[1] research findings suggested that live looping performance leans on artistic practices that used to be exclusively of music recording studios, eroding the boundaries

DOI: 10.4324/9781003154082-3

between studio and stage, and challenging our understandings of musical instruments and musical performance.

This chapter is organised into two main parts: the first will present an overview of the concept of instrumentality and a discussion of some features of *Import/Export*. The second part will discuss the performative research in the case study and will summarise with some reflections on the instrumentality of live looping devices and the correspondence between music performance in live and studio settings.

New instruments versus past concepts: reflections on instrumentality

Since the end of the 19th century, the technological developments in music production (e.g. electrification, recording and reproduction of sound, digitalisation) have contributed to the emergence of several modern and contemporary musical practices that came to challenge established definitions of musical instrument. On the one hand we have been observing an increasing variety of electronic and digital instruments, or even instruments that are strictly virtual or software-based; on the other hand, it seems that our understanding of the concept of musical instruments did not keep pace with these transformations, resulting in an incompatibility between past concepts and contemporary musical instruments and practices (Hardjowirogo 2017).

In the search for a concept that is inclusive towards contemporary musical instruments, Hardjowirogo (2017) argues that the usually accepted notion that musical instruments are "sound producing devices" (Hornbostel 1933; Lysloff & Matson 1985) seems no longer sufficient. Firstly, because musical instruments are not the only objects used to produce sound, and secondly, because instruments are more than just sound production devices: there are some objects that are immediately recognised as musical instruments, others that are not, and there are others *used* as instruments with a certain regularity (the saw played with a bow, an oil barrel used as a drum, among others). In conclusion, some objects can be more "instrumental" than others, and it would be possible to organise them according to their "degree of instrumentality" (Hardjowirogo 2017: 11).

That said, the concept of instrumentality emerges as an attempt to identify some transversal characteristics of musical instruments instead of presenting a universal notion, suitable for different musical practices, of what an instrument is. In short, instrumentality "denotes the potential for things to be used as musical instruments or, yet differently, their instrumental potential as such" (Hardjowirogo 2017: 17). For Hardjowirogo, instrumentality must not be interpreted as a property that an object essentially has or has not, but as a gradable, dynamic concept, which represents "a complex, culturally and temporally shaped structure of actions, knowledge, and meaning associated with things that can be used to produce sound" (Idem: 12).

The concept of instrumentality has been debated in recent years (Cance 2017; Hardjowirogo 2017; Baalman et al. 2018) and seems to have reached a critical point with the advent of digital instruments and the electrification and virtualisation processes of acoustic instruments (Alperson 2008; Kvifte 2008; Cance 2017; Enders 2017; Hardjowirogo 2017; Kim & Seifert 2017). What is particularly useful about this concept is that it takes into consideration aspects that are usually bypassed by past notions on musical instruments, for instance, the user's intention and the cultural context of the object/instrument. Cance, Genevois, and Dubois (2009) used conceptual and methodological tools from cognitive linguistics to better understand the notion of musical instruments on digital interfaces. Their study aimed to understand the concept of a musical instrument in contemporary musical practices and the findings showed that "instrument" not only refers to a particular device, but qualifies the nature of interaction between it and its users/performers. According to them,

> The instrumentality of these new devices, as well as of "classical" instruments, does not result from their intrinsic properties only. It is constructed through musical play, interactions between musicians and the design and development of the instruments, as illustrated by one user's comment: one is not born, but rather becomes, an instrument.
> *(Cance, Genevois, & Dubois 2009: 9)*

Thus, intention seems just as relevant as the material properties of the object itself, and this perspective shifts the focus from material features to immaterial ones, from collective traditions to individual volitions, and therefore it would be more appropriate to question *when* something is an instrument, instead of *what* an instrument is. Furthermore, musical instruments are embedded with an intangible repertoire of gestures and actions, which are at least as important as the user's intention and the physical attributes of the instrument, and, although it has been a central issue in the ethnomusicological study of musical instruments,[2] it seems to be disregarded in the study of digital contemporary musical instruments (Hardjowirogo 2017). Alperson (2008: 38) refers to it as the instrument's "immaterial features" and that these are commonly overlooked in the perception of musical instruments. According to him, it is necessary to look beyond the physical properties of musical instruments and consider their immaterial characteristics: "if we are to have a rich understanding of musical instruments, we cannot regard them simply as material objects. The moment they are musical instruments, they are musically, culturally, and conceptually situated objects" (Alperson 2008: 42). For Alperson our understanding of musical instruments shapes our way of thinking about music and the different ways music is produced and performed; above all, it plays a crucial role in how we appreciate many of the skills involved in music making.

By taking into account perspectives that are often neglected by other organological and musicological investigations, the concept of instrumentality is a useful theoretical framework to investigate the performance of live looping in *Import/Export: Percussion Suite for Global Junk*. As we will see next, this work features both electronic and acoustic instruments that deviate from the usual accepted notions of musical instruments (e.g. looper pedal and "found" objects). In addition, these instruments present singular sound, tactile, visual and ergonomic references to their performers, which may require non-standardised performance techniques and peculiar learning processes.

"Import/Export: Percussion Suite for Global Junk"

"Import/Export: Percussion Suite for Global Junk" (2008) by Gabriel Prokofiev (born 1975) is a solo work for percussion, live looping and live sound processing (hereafter LL and LSP respectively) featuring a quartet of "junk" objects as musical instruments: an oil drum, *Fanta*® bottles, plastic bags and a wooden pallet. Lasting about 30 minutes and organised into seven movements ("Voyages"), the work was commissioned by percussionist Joby Burgess and premiered in England in November 2008. Prokofiev explains that

> The seven voyages of *Import/Export* explore a quartet of truly global Junk objects. An Oil-drum, Wooden Pallet, Plastic Bag and Soda Bottle all travel the world – transporting, polluting and keeping our modern economy moving – the innocent yet destructive messengers of an ever hungry free-market. Using electronics to enhance the natural sounds of the metal, wooden, plastic and glass objects, the resulting music has a diverse range of textures, timbres composed with the traditional sense of structure, melody and harmony.
>
> *(Prokofiev 2008, programme notes)*

Notwithstanding the idea of found objects being generally associated with genres such as sound art, experimental or conceptual music, Prokofiev chose an opposed direction in *Import/Export*, creating "a pre-composed piece of music in the classical tradition that is 'musical' and has a sense of form, harmony & melody; not an experimental piece of conceptual sound-art" (Prokofiev 2008, Composer's notes). Some of its seven movements are structured similarly to the cyclical forms from the Western art music tradition (such as the *Rondo*, the *Scherzo*, or the *Sonata*), also presenting several indications of tempo, expressiveness and articulation that are common in this genre. The fusion of "junk" objects, LSP and LL, with musical elements and forms derived from the Western art music tradition contributes to the original character of *Import/Export*. According to Prokofiev,

> Electronics are used sparingly in this piece, with the aim of enhancing the natural sounds of the metal, glass, plastic and wooden objects. Whilst the lead part generally remains "clean", live looping is used to create multi-layered *ostinati* which are often processed.

While the LL is performed by the solo percussionist, the LSP should be performed by an extra electronic musician, who will deal with music scores specific for the audio processing, sound effects, audio panning, and mixing frequencies, as well as technical information on microphones and audio inputs/outputs/auxiliary channels. Having knowledge on LSP is not essential for the percussionist performer, but it might be helpful. For example, the interdependence of LL with music production may be expressed by its relationship with acoustics and microphones. In contrast to a sound recording studio, which is an acoustically controlled space, live music venues are usually much more unpredictable, presenting various types of reverberations, reflections, interference of external noises, and susceptible to audio feedback, among other. These conditions may make the LL and the LSP quite difficult to achieve, as percussionist Joby Burgess commented:

> The difficult always, with that piece [*Import/Export*], was putting it into a live show, because it is really hard to do that, the live processing. Well, just going to a big open space, put microphones on things and loop them, and trying to process them, and send signal back, it is just not easy. It is great if you are in a small, dry theatre, it works brilliantly, it is very easy to do many things. And it is really easy to do it in a recording studio, because there is no acoustic – which it's where Gabriel [Prokofiev] was doing everything, it is where we rehearsed everything. So as soon as we got our very first tour with that piece - think it was ten days running UK – and it was three, four, five venues, and some were pretty hard, the first one was pretty hard . . . so we had to find ways round that quite quickly, plus exploding *Fanta*® bottles in the first sound check, so rolling in touch for new pedals.
> *(Burgess in Bittencourt 2019)*

Percussionist Burgess also added that *Import/Export*

> is really difficult, it is time investment . . . I mean, there's nothing standard at all, is there? There is no "ah, I'll just go to the cupboard at college and get the oil drum out get all those objects out", or go get my looper. It is a big investment in time and, to some extent, money.
> *(Burgess in Bittencourt 2019)*

All these characteristics – unusual instrumentation, singular performance techniques, LL, LSP, work length (seven movements and about 30 minutes) – make *Import/Export* a complex, "demanding work" (Prokofiev in Bittencourt 2019).

What follows is an performative analysis of *Import/Export*, which includes musical excerpts in order to provide a more detailed understanding of the choices that guided the performance of the work.

Import/Export: a performer's perspective

Import/Export is organised into seven movements or "voyages" namely, *Voyage 1: On Land*; *Voyage 2: Engine Multiplier*; *Voyage 3: Cloudburst*; *Voyage 4: Plastic Invasion*; *Voyage 5 Fanta®*; *Voyage 6: Tropical Shores*; *Voyage 7: Memories of a Wooden Workhorse*. The instruments used are an oil drum (in Voyages 1, 2, and 6), plastic bags (in Voyages 3 and 4), *Fanta®* bottles (Voyage 5), and a wooden pallet (Voyage 7). These instruments present very unique sound characteristics and are performed through a variety of performance techniques, some adapted (from instruments such as marimba, symphonic bass drum or bongos, for example) and others that are work- or instrument-specific (for example, to create squeaky, rhythmic patterns through the friction of fingertips on a plastic bag, or to play some specific pitches of a scale with a soda bottle).

Bearing in mind that these instruments are everyday objects, it is essential that the performer has (or develops) an awareness of the different sound nuances they produce: "these objects are a central part of globalized world that we live, and their import/export movements keep our modern economy moving [. . .] But they can also create some beautiful sounds & harmonies, and under the hands of the *right percussionist* can make very exciting music" (Prokofiev 2008, notes for the performer, my italics). Prokofiev explained that a "right percussionist" is

> [. . .] a percussionist who . . . is curious and has an ear to really look, to really try to find the best sound that an instrument can offer [. . .] some musicians are technically brilliant, but sometimes you can feel in their playing, the notes . . . they're doing everything but they're not . . . really trying to get the very best tone. [. . .] So I think some percussionists might be brilliant in rhythm but they might not be so focused on the nuances, on the details of the sound. So I think when you are using unusual objects you have to be extra curious to really try to find the right sound [. . .] You have to be open minded. [. . .] It has to do with a percussionist . . . who really cares about sound and tries to find it. [. . .] Specially when it is not a proper percussion instrument you have to work a bit harder to get it to really sing, you do have to work a little bit harder

to really bring out the sound. And the amplification can help because equalisation particularly . . . you can bring out some of the more musical frequencies, and that can enhance the sound. [. . .] But is also how you hit it and I think some musicians might not be patient enough to really try to find the sound.

(Prokofiev in Bittencourt 2019)

The materials of each instrument of *Import/Export* (metal, plastic, glass, and wood) present very distinct sound qualities: for example, the oil drum can produce long, resonant and sustained sounds; by contrast, the plastic bags have a much softer character, producing sounds of low volume; the glass bottles are surprisingly pitched, and can also create "marble" sound textures; and finally, the wooden pallet has a prominent staccato, dry sound quality, but it also has nine distinct pitches according to each of its slats. An awareness of the ways that sound may be produced, and of the sonic characteristics of these instruments, is directly related to live audio recording and the attempt to get the best tone in each recording take, and therefore also impacts the performance of LL.

In *Plastic Invasion* (Voyage 4) for example, the performer must approach amplification carefully while performing the plastic bags, since the dynamics mostly depend on the distance between the bags and the microphone. As some performance techniques produce contrasting levels of volume (the "squeak" technique[3] is considerably louder than the "fist squeeze",[4] for instance), it is expected that the performer will play with the distance between the mics and the sound source, to balance the dynamics of each technique.

Performing live looping

The use of LL is an essential feature in the performance of *Import/Export* and therefore the know-how about looper devices must be mastered by the performer. The work was conceived as a multitrack piece, using LL to create and reproduce different sound layers simultaneously. Each of the seven movements is notated with indications of musical content to be recorded and reproduced.

The LL technology originally used in *Import/Export* was based on hardware: a looper pedal with three independent tracks or channels to record, reproduce and overdub layers of sound. In 2008 when the work was composed, hardware-based LL technologies were very popular and this was also adopted for its world premiere by percussionist Joby Burgess. Similarly, I opted to use a looper pedal in my first performances of *Import/Export*, including its Portuguese premiere.[5] As with any instrument, several hours of investigation, practice and experimentation were necessary to explore its instrumental potential. Indeed, it was not just a matter of studying the score of *Import/Export* and getting to know the looper, but also of understanding its instrumentality within the context of the work.

As a performer, one of the reservations I have regarding LL concerns the predictability of how the performances usually unfold, which consequently brings some invariability to the compositional form or musical architecture. Pachet et al. (2013) commented that since loop pedals systematically playback the recorded content without any variation, it can make performances monotonous and boring, both to the musician and the audience, producing a canned music (pre-recorded music) effect[6] (Pachet et al. 2013: 2205).

However, in *Import/Export*, the looper pedal is used to achieve distinct goals and not exhaustively: for example, in *On Land* (Voyage 1), it is discreetly used in much of the piece, mainly for the recording of a melodic line (in a single track and without overdubs), which will be the basis for a solo by the percussionist; in *Cloudburst* (Voyage 3) the looper acts more as a textural instrument, to create a sonic environment similar to "a synthetic storm" (Prokofiev 2008); in *Engine Multiplier* (Voyage 2) the looper is explored throughout the whole piece to produce a dense, multi-layered rhythmic texture like "a menacing inexorable engine" (Prokofiev 2008, 1); and in *Memories of a wooden workhorse* (Voyage 7) the looper is used only for the second half of the piece to create melodic counterpoints and rhythmic *ostinati* of an almost dance-like character.

Technically, the performance of *Import/Export* requires considerable skill to manage all the demands and choreography necessary to perform the instruments while simultaneously recording, reproducing, overdubbing, and editing audio loops. In *Engine Multiplier* (Voyage 2) the performer needs some agility to deal with the short time available to activate and deactivate the recording and playback controls of the loops, while simultaneously changing drumsticks and mallets. For example, in Excerpt 1[7] the performer must interrupt the playback of three soundtracks (represented by the horizontal numbered lines below the staff) and at the same time start recording new musical content on Track 1 (rectangle above the staff):

FIGURE 1.1   Ex. 1 – Voyage #2: Engine Multiplier, bars 26–28

(Prokofiev 2008, 2)

Excerpt from the original digital score.

This excerpt in particular is more complex to perform than its notation indicates, since there are some hidden procedures to be made. In order to record the new musical content (Loop 1), firstly it is necessary to empty Track 1 in the looper by erasing the earlier music content already stored on it. However, to delete the content of a track from the looper, it is necessary to press and hold the "stop" button[8] of Track 1 for two seconds, that is, it is not an instantaneous procedure. My strategy was to connect an external footswitch in the looper to expand the possibilities to control the device. Thus, by assigning an additional "erase" function to the footswitch, I was able to empty Track 1 slightly before the starting point of recording new audio (bar 27), and then simultaneously press the "stop" and "recording" buttons to interrupt all track/loops and start a new recording on Track 1. This is just one example of occult performance actions involved in the LL of *Import/Export*. There are several situations similar to this excerpt throughout the work, and the performer must study the potential of the looper pedal – its instrumentality – as well as seek creative solutions regarding eventual inconsistencies between notation and performance.

As any musical work, the performance of *Import/Export* also includes some choreography: a repertoire of gestures and movements necessary to perform it. However, in musical works such as this, which requires multiple instruments, drumsticks, mallets and implements, in addition to LL, the fluidity of the performance can be easily affected. Percussionist Joby Burgess commented on this:

> I think the thing with that piece [*Import/Export*] is getting it to flow. Because all of those pieces where you are playing stuff, and you've got some . . . for example, the oil drum, you've got two dozen different implements to play with, and you've got all these different techniques, which comes from playing tabla, playing congas and djembes, and all these different sorts of mallets, and scrapers and things . . . so is getting those, and is getting the flow of it, because sometimes you do a little thing and it goes into the mix then you've got play the next thing. So it is like all the pieces, making all these gestures, you're doing part of it, as opposed to "oh I didn't do that right I've got to do this", and following a score at the same time. [. . .] So it's a different ball game . . . so I'm not playing one instrument and then doing the electronics, I'm doing that and then I have to think about this, and then do this. So it's the sort of . . . I suppose the choreography of it was bigger. So if we think about . . . let's say, *Zyklus*, by Stockhausen: if you play that, there is this massive choreography either. You know, how do you get from the log drum there to cowbell here and playing a triangle over there, you know? The tam tam behind you, with the right object. There is all those challenges but then with Gabriel's piece there is that plus you have to do the looping. And if you don't do the

looping then the piece doesn't carry on, so . . . Yeah, it was getting that to really flow as a piece, I think that was the big challenge.

(Burgess in Bittencourt 2019)

Notably, some choreographic fluidity is required for the performance of *Tropical Shores* (Voyage 6), in which some audio loops must be activated and deactivated constantly, and in short intervals of time. Excerpt 2 demands a series of precise movements of the feet for the looper pedal, in addition to the performance of the oil drum. In this excerpt, the percussionist must perform the solo on the reproduction of the first loop (Loop 1), while Loop 2 is constantly activated and deactivated (indicated by the text "punch loop" below the stave):

FIGURE 1.2 Ex. 2 – 43 Voyage #6: Tropical Shores, bars 41–43

(Prokofiev 2008: 2)

Excerpt from the original digital score.

In this excerpt, some precision is required to turn the Loop 2 on and off at exact points in time, and an agile tactile sense with feet as well, in order to avoid looking at the pedal every time to press the buttons, which would compromise the fluidity of the performance. During my study sessions, I used to practise this section with bare feet or socks, to increase my sensitivity and touch and perform the choreographic movements of the looper in an automated way.

The whys and whens of live looping

My performance of *Import/Export* has evolved in several aspects since I started to work on it – especially regarding the use of LL. From the very beginning of my work on the piece, some of my reflections have been focused on to what extent the LL influences the performativity of the work, and on the possibilities and limitations of the ever-changing LL technologies. One aspect in particular has to do with the fact that LL performance (in the case of acoustic instruments) is extremely entangled with, or even dependent on, other factors such as microphones, amplification and, above all, the acoustics of the performance space. In addition, the technological apparatus available for LL performance has been improved since 2008 (when *Import/Export* was composed), especially with regard to software: currently, almost any DAW[9] incorporates built-in or third-party plugins dedicated to LL, allowing the performer to have full control over recording, playback and looping, among other features.

Unfortunately, in my first performances of *Import/Export* I could not obtain the desired sound results due to audio feedback problems, which resulted from the acoustics of some concert halls having excess reverberation or similar issues. After reflecting on these issues, I decided to move from hardware to software: instead of using looper pedals, I started using the computer connected to a MIDI pedalboard, which allowed the performance of LL and also the use of audio effects (such as equalisers, envelopers and filters) to shape the sound. However, this was not enough to avoid feedback issues and the excessiveness of "room sound" in the recorded loops. The sound quality had definitely improved, but the acoustics of some performance spaces kept causing audio feedback issues.

At some point in the midst of a Portuguese tour featuring *Import/Export*, I questioned myself why should the musical contents of the piece be recorded during each performance (and at each concert venue) if they are fixed, similar to the tape part of electroacoustic works? This question led me to the second (and perhaps the most significant) change, which was to use pre-recorded audio loops, instead of recording them live. Therefore, most audio tracks, which used to be recorded and looped during a live performance, were previously recorded in the studio to ensure good sound quality and to preclude audio feedback issues. This change has produced the best results so far, and has also raised some personal reflections on LL and the notion of liveness that will be discussed subsequently. With this new setup the performance of *Import/Export* remains unchanged in terms of musical narrative – the only difference is that the audio loops are recorded beforehand to safeguard their sound quality. In other words, the "recordivity" (Knowles and Hewitt 2012) was bypassed but the performativity is still present – it is worth remembering that live looping is at the service of the performance, but it is not the performance.

Consequently the "live looping" became looping, since the live recording task was excluded from the majority of the work (but not completely). Importantly, these changes were not an attempt to simulate the performance or to make it easier; instead, they aimed to enhance the sound quality of the instruments and focus in the music performance as a whole, instead of having the act of recording as a central feature. Especially with musical works that feature everyday objects as instruments, sound quality is crucial (Bittencourt 2019). Therefore all the sonic, harmonic, melodic and rhythmic characters of the work were improved, the audience could experience the work better and the performance has progressed.

However, not all parts of *Import/Export* could be recorded beforehand, since there is some instrumental unpredictability involved in the work. An example of this is the end of *Fanta*® (Voyage 5), in which the glass bottle must be used as a wind instrument to create a polyphonic texture with the looper pedal. The excerpt below (Ex. 3) illustrates the recording and playback of the first of the loops:

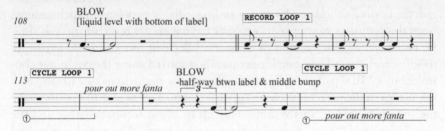

FIGURE 1.3   Ex. 3 – Voyage #5: Fanta®, bars 108–118

(Prokofiev 2008: 4)

Excerpt from the original digital score.

The different pitches of this polyphony are generated from the manipulation of the amount of liquid contained in the bottle: before recording each loop or overdub, the percussionist must gradually pour out small amounts of liquid from the bottle, which enables the production of different pitches (ultimately, the smaller the amount of liquid in the bottle, the lower the notes). Since the amount of poured liquid is not accurate, it brings a somewhat random character to the resulting pitches, so different notes might be produced in each performance. Having this in mind, I decided to keep the live recording of the loops for this section, so as to avoid possible discrepancies of pitch.

As previously mentioned, my performative experience with *Import/Export* triggered some reflections on the possibilities and limitations of LL performance, especially in the context of notated Western art music. As *Import/Export* is completely notated, the role of LL within this work is seemly different than it usually plays as a prominent technique among solo performers in the contexts of pop, jazz and improvised music. In the next section I will present some arguments to demonstrate the relationships of LL performance with real-time, improvised music creations, and studio practices, in addition to a reflection on the skills associated with LL, how they are perceived by the audience, and how audience perception might influence the instrumentality of digital musical instruments such as loopers.

From studio to stage: blurred lines

The use of digital technologies in live performances has progressively reduced the boundaries that separate the musical practices of recording studios from those of live concerts (Knowles and Hewitt 2012; Kjus and Danielsen 2016; Renzo and Collins 2017). Digital devices for LL are examples of such technologies, available both as hardware (standalone units, pedals) or software (plugins or computer applications). Live looping can be defined as "the simulation of multitracking studio techniques in 'real time'" (Richardson 2009: 89). Despite that LL digital devices became more accessible to the general public from the 2000s onwards, the looping technique is older than that and rooted in analogue technologies, such as tape loops with magnetic tape and locked grooves in recorded discs, created by pioneer composers such Pierre Schaefer, Paul Hindemith, Karlheinz Stockhausen, Pauline Oliveros, and Terry Riley, among others (Holmes 2008 [1985]; Renzo and Collins 2017).

Notably, these digital technologies for music – also referred to as "threshold technologies" (Knowles and Hewitt 2012: 6) – not only blurred the line between studio and live performances but have also induced performing musicians to develop skills that are more frequently required in music production studios (such as recording, mixing, and editing, among others). Through digital music instruments (hereafter DMIs) these techniques, which are generally conducted in private settings and in distinct periods of time, started to be integrated into the performative act by the musicians on stage, in real time. A series of implications has emerged with the live performance with DMIs, ranging from questions on the instrumental potential of these devices, the different notions of virtuosity that they may encompass, to reflections on how their inherent skills and performance modes are perceived by the public (or even if they really are). Particularly with regard to the audience's perception of the technical demands and expertise involved in LL, Knowles and Hewitt argued that

> Contrary to what one might expect, recording on stage often has a heightened sense of liveness. The experience of liveness increases as the inputs are staged and made clear to the audience. During the loop record cycle the audience sees the performer attempt to perform a flawless take, knowing that any errors would have serious consequences as there is no simple way to erase an error or re-do a take in the middle of a performance. It illustrates recording as a performative act not just in terms of the performer in a traditional instrumental sense but also the recording engineer/producer and their playing of the technological instruments of recording.
>
> *(Knowles and Hewitt 2012: 20)*

Notwithstanding the fact that an audience may have some grasp of what is happening on the stage, we cannot assume that they possess the expertise to really understand the demands involved in a LL performance, nor that they are aware that any errors would have serious consequences or even ruin a performance, as suggested by Knowles and Hewitt (2012). In fact, the situation seems to be the opposite, as Gracyk explains: "if one does not know the demands of the particular instrument, one cannot judge the virtuosity displayed. And this may be the situation more often than not" (Gracyk 1997: 145). In this argument, the expression "particular instrument" here could be applied to LL instruments and DMIs without imbroglio. As the concept of instrumentality demonstrated, the word "instrument" provides room for several connotations and interpretations, especially because "it appears that 'instrument' does not actually refer to a device [. . .] but rather qualifies its interaction with users [. . .]" (Cance et al. 2013: 297), and the actions and meanings embedded in an instrument probably deserve more attention than its physical attributes. Regarding the instrumental potential of LL devices, I asked percussionist Joby Burgess about his viewpoint on the looper pedal in the context of *Import/Export*:

And how do you see the looper pedal in the [instrumental] *setup?*
It is a kind of pretty important! It is a kind of the big thing and the small thing. Because without it the piece doesn't go anywhere, because it is the multitrack thing. So, to begin with it, it is one of those things where the audience goes "wow! He's looping that". But as soon as you get the third or fourth piece, they are not thinking about the looper, it is just part of how the sound is coming to be, and how it has been born, so . . . How do I think about it? I don't know, I'm not sure I think about it . . . I don't know . . . I was about to say that I think about it as an instrument, but maybe because it is not really an instrument . . .

Depends on your concept of instrument.
Yeah . . . but it does not make a sound, does it? So . . . but I do have to perform, I have to play it. Yeah, maybe it is an instrument, I don't know . . . I don't know, to be honest. It is part of the kit. If I forgot it, it would be a big problem, I mean, if I forget a few sticks that's fine, but if I forget the looper, forget it, no gig!

(Burgess in Bittencourt 2019)

As theoretical perspectives on instrumentality have demonstrated, to produce sound is undoubtedly an essential feature of musical instruments, but there are more to be considered, especially regarding contemporary music instruments. For example, audience perception, which has to do with meeting the audience's expectations about the demands and degree of challenge involved in the performance, and how these are perceived. As Hardjowirogo has argued, "instrumentality, in the sense of a category that legitimates instrumental performance, is highly dependent on audience perception" (2017: 21). In turn, the audience's perception of instrumentality seems to be fundamentally connected to the sense of "liveness", a concept exhaustively discussed by Auslander ([1999] 2008) built primarily around the audience's affective experience. For Auslander "liveness" is not a concept defined in ontological terms, but a historically variable effect of "mediatization", since "the idea of what counts culturally as live experience changes over time in relation to technological change" (Auslander 2012: 3). The understanding of what a musical instrument is at this digitally-driven era – a time in which virtual/interactive/intelligent/automated resources, mediums, systems, and interfaces are increasingly eroding the boundaries of past concepts on the subject – is still lacking. Similarly, the audience perception and experience of liveness concerning DMIs seem to still be under development, at a slower pace than the understanding of their instrumental potential.

As defended by Knowles and Hewitt elsewhere in this text, "the experience of liveness increases as the inputs are staged and made clear to the audience" (2012: 20), which suggests that the relationship between input (the performer's actions) and output (resulting sounds) is crucial for the audience's perceptual experience of liveness and of the instrumentality of a device/instrument. However, DMIs rarely present a clear causal relationship on how a performer's gestures and actions are transformed in sounds – an issue also referred to as "gesture-sound causality" (Emerson and Egermann 2017: 364). Therefore, obscure relationships of "gesture-sound causality" become a struggle for the audience when experiencing new instruments and their instrumental potential, as inputs and outputs seem to be disconnected from each other. The singer, composer, instrumentalist and music producer Imogen Heap, who frequently uses DMIs and LL techniques to create spontaneous compositions during her live performances, commented about this issue: "Because I'm sort of barricaded by this gear I'm sort of like the Starship Enterprise. I don't think that people in the audience can actually see what's going on. They can see my hands moving but they don't really know what I'm doing" (Heap[10] in Knowles and Hewitt 2012).

Furthermore, in a reflection on how the audience experiences a musical performance, Alperson argues that there is a double awareness or, using his own words, a "twofoldness" of musical performances: that is, the audience's double consciousness of the performance *of the* work and the performance *in the* work. In spite of his belief that the audience may have some consciousness of this sort of duality when appreciating the performance of most musical works, Alperson explains that not every listener is equally capable of appreciating all the nuances of it, and

> With respect to the actual demands placed upon musical performances, it is possible that performing musicians are themselves most likely to have the fullest appreciation of what has been accomplished, since they are most familiar with the demands of the project from the inside, so to speak. In fact, it is likely that there is a natural segregation of listeners in this regard. French horn players are in a better position to understand the level of achievement of French horn playing than non-musicians or even than

musicians who do not play the French horn. Indeed, very good French horn players are likely to be in a better position to understand the level of achievement of very good French horn players than mediocre French horn players.

(Alperson 2008: 48)

Alperson's argument reveals a distinct facet of audience perception, focused on how music performances are perceived not by general listeners but performing musicians themselves (that is, specialised listeners). If a specialised listener may be unable to fully understand the degree of accomplishment and demands of a musical performance, on a standard instrument, by a peer, what we could say about the performance of DMIs, which involves extrinsic instruments with a less evident (and sometimes incomprehensible) instrumentality? Here "instrumentality" is used as an umbrella term that embraces a whole package of intangible or "immaterial features" (Alperson 2008) – repertoires of actions, gestures, skills, performance techniques, sonic aesthetics, learning procedures, and unfamiliar notions of musical virtuosity, among others. In the performance of loopers and DMIs in general, the (expected) unfamiliarity of the audience with the technological apparatus might blur their focus between the performance *of* the work and *in* the work.

Final reflections

Undoubtedly, the use of live looping resources expands the potential of live music performances, making possible, for example, the public presentation of musical creations conceived exclusively in recording studios and not intended for live settings. However, live looping devices and digital musical resources also influence music performances at different levels, especially in the way such devices are perceived by the audience, the relationship between gesture and sound production, and their sense and experience of liveness. In addition, these devices and resources entail deeper epistemological and ontological questions on musical instruments, instrumentalists, and performance (for example, on the type of knowledge and skills involved in performing these resources, on the peculiar notions of instrumental virtuosity that they may evoke, the corporeal interrelationships between instrument, habit and technique, and our conceptions of what is or may become a musical instrument, among others).

Importantly, the understanding of live looping devices can be refined from the theoretical perspectives on the instrumentality concept. As mentioned earlier, digital and acoustic musical instruments differ in many ways, and even a fundamental criterion for any musical instrument (such as to produce sound in some way) may be radically different (for example, with digital musical instruments, sound does not come from the physical and acoustic properties of the material object itself, as is the case with standard acoustic instruments). There are many other criteria[11] to be considered in the perception of musical instruments, and it is perhaps more coherent to consider the idea proposed by Hardjowirogo (2017), that some objects can be more "instrumental" than others, or that some can have an instrumental potential in a particular situation but not in others. Furthermore, as suggested by Alperson (2008), the understanding of musical instruments also shapes our understanding about music and the distinct ways in which it is produced and performed, and, above all, it also plays a crucial role in how we appreciate many of the skills involved in music making. Thus, it seemed superficial to investigate the demands and particularities of the performance with live looping instruments without first reflecting on their instrumentality, through distinct theoretical and artistic perspectives on musical instruments in contemporaneity.

In an attempt to broaden the understanding of live looping and digital music instruments, it is also important to take a closer look at the potential roles of technology in connecting intentions, actions and sound production (Kjus and Danielsen 2016). Particularly in the use of the "threshold technologies" (Knowles and Hewitt 2012: 6), there seems to be a certain consensus on the necessity of making performative actions clear to the spectators and, above all, balancing the relationship between gesture and sound. Knowles and Hewitt defended that "the use of the looper pedal is foregrounded in performance and thus makes the process of recording a point of significant focus within the performance itself" (Knowles and Hewitt 2012: 13) – in short, "recordivity" becomes the performance, or part of it. On the other hand, Croft has noted "if the relationship between the energetic and gestural characteristics of the performer's action and the sound generated is opaque, then most of the point of live performance is lost" (Croft 2007: 61).

Hence the musician's actions to record the loops live become performative elements in themselves and, at the same time, it is important that these actions present axiomatic causal relationships between gesture and the resulting sound. In my case with the performance of *Import/Export*, a technical aspect of live looping (namely, the unpredictable acoustics of concert venues) was compromising the sound quality and bedevilling the performance as whole. The awareness of the role of technology and its connection with my artistic intentions led me to adjust or balance some processes of my performance, and the degree of "recordivity" was reduced so that other dimensions of performativity could gain prominence.

Through the utilisation of LL in *Import/Export* the percussionist's set of skills is expanded to encompass knowledge and techniques from the audio recording/production domain, and reinforces the intertwining of LL instruments with music recording studios and their artistic practices. In *Import/Export*, Western art music and popular music are synthesised within a singular artistic creation, and its singularity is shaped by the tensions that may arise from the intersection of these two apparently contrasting musical domains (e.g. the transmission of musical ideas through musical scores versus oral/ear tradition, public live performances versus music recordings as the primary activity of performers, the predominance of fully notated compositions versus musical creations that incorporate (or are based) on improvisational practices, among others). In the understanding of a musical instrument, the cultural context in which it is inserted cannot be neglected, and, notwithstanding that different scholars have drawn attention to this subject (Kartomi 1981; Racy 1994; Dawe 2002; Alperson 2008; Bates 2012; Cance 2017), the relationship between musical instruments and culture is still often overlooked in the study of contemporary musical instruments (Alperson 2008; Hardjowirogo 2017). Importantly, the comprehension of live looping devices as musical instruments (their instrumentality) must undoubtedly consider the artistic and cultural contexts in which they have been rooted.

In the encounter between Western art music and popular/urban music presented in *Import/Export*, some artistic skills of the latter may be not recognised as such (or, at least, expected) in the former. Particularly with regard to live looping, these skills include how musical ideas are recorded, structured and arranged in real time, the flow of musical sections/improvisations, the interaction with the recorded loops, the ability to mix layers of sound and blend audio effects, and the act of composing in front of the audience (composition as performance). On the other hand, the performative settings usually presented in the performance of Western art music may be seen to be somewhat dull, or lacking energy in some way, by those accustomed to pop music concerts. In contrast to acoustic musical instruments and their long tradition of historically habituated modes of performance and presentation of works, it seems that new digital musical instruments such as loopers are in the midst of an ongoing process towards the construction of their instrumentality, a process in which audience perception, of the demands involved in musical performances, and the cultural embeddedness of musical instruments play a fundamental role.

Notes

1 Performer in charge of the commission and world premiere of the work.
2 For more on this subject, please see Kartomi 1990.
3 "A small section of the bag is stretched between the thumbs of both hands. This tight stretched area is then rubbed with fingers, using friction to create squeaks" (Prokofiev 2008).
4 "The Plastic Bag is squashed up inside a fist. The fist is held gently closed & still so that no sound is made. To make a noise the fist is quickly squeezed" (Prokofiev 2008).
5 Presented at the Convento de São Francisco, Coimbra, Portugal, on May 16th, 2018.

6 "Canned music, piped music (terms almost always used with negative connotations of the mechanical)" (Frith 2003, 151).
7 All score excerpts with the kind permission of the composer and his publisher, Mute Song.
8 This is the procedure to erase a track in the RC-300 Boss Loop Station®, which may be similar for other looper devices.
9 Digital Audio Workstation.
10 Heap, Imogen. 2011. *Interviewed by Melody Alderman*. Pure Songwriters. Available at http://www.puresongwriters.com/artists/imogen_heap.html#.TzB52vGn3dU and consulted on August 28th, 2021.
11 Hardjowirogo (2017) has defined a number of criteria that, based on a literature review, appear to be crucial for the construction of instrumentality: 1) Sound production; 2) Intention/purpose; 3) Learnability/Virtuosity; 4) Playability/Control/Immediacy/Agency/Interaction; 5) Expressivity/Effort/Corporeality; 6) "Immaterial Features"/Cultural Embeddedness; 7) Audience Perception/Liveness.

References

Alperson, Philip. "The Instrumentality of Music." *Journal of Aesthetics & Art Criticism* 66, no. 1 (2008): 37–51. doi: 10.1111/j.1540–594X.2008.00286.x.

Auslander, Philip. *Liveness: performance in a mediatized culture*. 2nd ed. (London: Routledge, 2008).

Auslander, Philip. "Digital Liveness: A Historico-Philosophical Perspective." *PAJ: A Journal of Performance and Art* 34, no. 3 (2012): 3–11.

Baalman, Marije, Simon Emmerson, & Oyvind Brandtsegg. "Instrumentality, perception and listening in crossadaptive performance." ICLI 2018, 4th International Conference on Live Interfaces. Inspiration, Performance, Emancipation, Porto, Portugal (2018).

Bates, Eliot. "The social life of musical instruments." *Ethnomusicology* 56, no. 3 (2012): 363–395.

Bittencourt, Luís. 2019. "Percussão e instrumentalidade: explorando a performance de instrumentos e fontes sonoras incomuns." Doctoral thesis in Music, Departamento de Comunicação e Arte, Universidade de Aveiro.

Bittencourt, Luís. "Percussion and instrumentality: exploring the performance of unusual instruments and sound sources." In *Hidden Archives, Hidden Practices: Debates about Music-Making*, edited by Helena Marinho, Maria do Rosário Pestana, Maria José Artiaga and Rui Penha (Aveiro: UA Editora – Universidade de Aveiro, 2020).

Cance, Caroline. "From Musical Instruments as Ontological Entities to Instrumental Quality: A Linguistic Exploration of Musical Instrumentality in the Digital Era." In *Musical Instruments in the 21st Century: Identities, Configurations, Practices*, edited by Alberto de Campo Till Bovermann, Hauke Egermann, Sarah-Indriyati Hardjowirogo and Stefan Weinzier, 25–43 (Singapore: Springer, 2017).

Cance, Caroline, Hugues Genevois & Daniéle Dubois. "What is instrumentality in new digital musical devices? A contribution from cognitive linguistics and psychology." *ArXiv* abs/0911.1288 (2009): n.p.

Cance, Caroline, Hugues Genevois, & Danièle Dubois. "What is instrumentality in new digital musical devices? A contribution from cognitive linguistics &

psychology." In *La Musique et ses instruments*, edited by Michèlle Castellengo & Hugues Genevois, 283–297 (Paris: Delatour, 2013).

Croft, John. "Theses on liveness." *Organised Sound* 12, no. 1 (2007)): 59–66.

Dasgupta, Subrata. "Technology and Complexity." *Philosophica* 59 (1997): 113–139.

Dawe, Kevin. "The Cultural Study of Musical Instruments." In *The Cultural Study of Music: an Introduction*, edited by Trevor Herbert & Richard Middleton Martin Clayton. (London and New York: Routledge, 2002).

de Assis, Paulo. *Logic of Experimentation: Reshaping Music Performance in and through Artistic Research* (Leuven: Leuven University Press, 2018).

Emerson, Gina, & Hauke Egermann. "Mapping, Causality and the Perception of Instrumentality: Theoretical and Empirical Approaches to the Audience's Experience of Digital Musical Instruments." In *Musical Instruments in the 21st Century* (Singapore: Springer, 2017), 363–370.

Enders, Bernd. "From Idiophone to Touchpad. The Technological Development to the Virtual Musical Instrument." In *Musical Instruments in the 21st Century*, (Singapore: Springer, 2017), 45–58.

Frith, Simon. "Music and everyday life," in Martin, Clayton, Trevor Herbert & Richard Middleton (Eds.). *The Cultural Study of Music: A Critical Introduction* (New York & London: Routledge, 2003), 92–101.

Gracyk, Theodore. "Listening to Music: Performances and Recordings." *The Journal of Aesthetics and Art Criticism* 55, no. 2 (1997): 139–150.

Hardjowirogo, Sarah-Indriyati. "Instrumentality. On the Construction of Instrumental Identity." In *Musical Instruments in the 21st Century* (Singapore: Springer, 2017), 9–24.

Haseman, Brad. "A manifesto for performative research." *Media International Australia, Incorporating Culture & Policy* 118 (2006): 98–106.

Holmes, Thom. *Electronic and experimental music: technology, music and culture* (London and New York: Routledge, 2008).

Hornbostel, Erich Moriz Von. "The Ethnology of African Sound-Instruments. Comments on Geist und Werden der Musikinstrumente by C. Sachs." *Africa* 6, no. 2 (1933): 129–157. doi: 10.2307/1155180.

Kartomi, Margaret J. "The processes and results of musical culture contact: A discussion of terminology and concepts." *University of Illinois Press on behalf of Society for Ethnomusicology* 25, no. 2 (1981): 227–249.

Kartomi, Margaret J. *On concepts and classifications of musical instruments.* (Chicago and London: University of Chicago Press, 1990).

Kim, Jin Hyun, & Uwe Seifert. "Interactivity of Digital Musical Instruments: Implications of Classifying Musical Instruments on Basic Music Research." In *Musical Instruments in the 21st Century* (Singapore: Springe, 2017), 79–94.

Kjus, Yngvar, & Anne Danielsen. "Live mediation: Performing concerts using studio technology." *Popular Music* 35, no. 3 (2016): 320–337.

Knowles, Julian, & Donna Hewitt. "Performance recordivity: studio music in a live context." Proceedings of the 7th Art of Record Production Conference 2012, San Francisco State University, San Francisco, CA.

Kvifte, Tellef. "What is a musical instrument." *Svensk tidskrift för musikforskning* 1 (2008): 45–56.

Lysloff, René T. A., & Jim Matson. "A New Approach to the Classification of Sound-Producing Instruments." *Ethnomusicology* 29, no. 2 (1985): 213–236. doi: 10.2307/852139.

Pachet, François, Pierre Roy, Julian Moreira, & Mark d'Inverno. "Reflexive loopers for solo musical improvisation." Proceedings of the SIGCHI Conference on Human Factors in Computing Systems, Paris, France, 2013.

Prokofiev, Gabriel. 2008. Import/Export: Percussion Suite for Global Junk. **Author's edition.**

Racy, Ali Jihad. "A dialectical perspective on musical instruments: The East-Mediterranean mijwiz." *Ethnomusicology* 38, no. 1 (1994): 37–57.

Renzo, Adrian, & Steve Collins. "Technologically Mediated Transparency in Music Production." *Popular Music and Society* 40, no. 4 (2017): 406–421. doi: 10.1080/03007766.2015.1121643.

2

HASGS

Its repertoire using live looping

Henrique Portovedo

Introduction[1]

The invention of musical instruments serves the purpose of increasing the expressive capacities of human beings, and each instrument is, in itself, a technological example of its time. For example, the need for greater sound projection and volume arose during the 19th and 20th centuries, resulting in the reconfiguration of instruments (e.g., changing their size and shape). These modifications, however, did not alter the fundamental nature of these acoustic instruments, and the characteristics that define them were largely maintained. The first experiments with recorded and synthesised sounds, carried out by Léon-Scott, Edison, Helmholtz and others (Collins, Schedel, and Wilson 2013) in the 19th century, allowed the realisation of works that combined acoustic, mechanical and electronic elements. The emergence of electricity led to the development of revolutionary and influential new devices, such as microphones and loudspeakers, that created the capacity for amplification, recording, and musical reproduction. More recent advances in technology have revealed countless ways to endow instruments (Miranda & Wanderley 2006) with new performative possibilities.

In recent years, the proliferation of new digital instruments (DMIs) has been enormous, together with the recent resurgence of electronic instruments with mixed characteristics. Examples include analogue-hybrid and digital-analogue synthesisers that can be modular, include a keyboard, or be controlled by wind. Wind-controlled synths have performative interfaces similar to those of wind instruments, not only in terms of fingering, but also in terms of articulation, dynamics, amplitude and other physical characteristics. This is especially noteworthy given that many instrumentalists and even composers no longer need to develop a virtuosic or highly developed piano technique, moving them away from electronic exploration via the keyboard-based interface model. In response to the proliferation of electronic technology associated with new and hybrid musical instruments, the terms extended (Penny 2009; Normark, Parnes, Ek, & Andersson 2016) augmented (Schiesser & Schacher 2012: Thibodeau, & Wanderley 2013: Kimura, Rasamimanana, Bevilacqua, Zamborlin, Schnell, & Fléty 2012) and prefixes such as hyper- (Machover 1989); (Palacio-Quintin 2003), meta- (Impett 1994; Burtner 2002), infra- (Bowers & Archer 2005), and even mutant- (Neill & Jones 2010), were coined to emphasise the different approaches chosen.

In the case of the work presented here, regarding the Hybrid Augmented Saxophone of Gestural Symbiosis (HASGS), as referred to in the terminology itself, it was decided to keep the term Augmented, since the augmentation process occurs in a non-intrusive or destructive way, in relation to the mechanics of the acoustic instrument, trying to add characteristics without compromising its organic qualities, and maintaining the qualities that define the instrument, in terms of both technical and sonic characteristics. In this sense, the IRCAM[2] Augmented Violin Project (Kimura, et al. 2012), the Magnetic Resonator Piano (McPherson 2010) or, more recently, SABRE by Matthias Mueller appear to be more aligned with the philosophy developed here.

HASGS

The Hybrid Augmented Saxophone of Gestural Symbiosis was initially developed with a DIY ("do it yourself") logic by myself as a virtuoso classical saxophonist and it was largely justified by some of the repertoire that I was performing and premiering, including contemporary repertoire for saxophone and electronics. The same approach was taken during the construction of the initial phase of prototypes, taking into consideration the challenges posed by new works that could be written for the system.

While the initial concept for the HASGS was to develop a system that understands the functionality of an Electronic Wind Instrument (EWI) integrated into an acoustic instrument, this idea was soon abandoned because the instrument's physical structure included an excessive number of sensors. This initial approach posed problems, not only with the instrument's structure and ergonomics, but also because it provided myriad technical possibilities that might jeopardise the virtuosity developed over more than 20 years of instrumental practice. Therefore, the term Reduced Augmentation can be applied here, since the technology in the new instrument is aimed at significantly boosting performance parameters and not constraining them, thereby lessening the impact on the existing affordances of the acoustic instrument.

In the NIME context, one of the fundamental problems regarding the proliferation of augmented instruments is that they are short-lived. This lack of longevity is due to several factors, including an absence of written repertoire by a community of composers and the fact that these instruments are almost exclusively performed by their creators. Another fact is that most performers who use these instruments view improvised aesthetics as a way of making musical expression as free as possible. Thus, an important point of discussion for the NIME community to consider is the development of communities around the instruments to help performers to explore the full potential of these instruments. Such communities may be able to contribute to the proliferation of repertoire and to the iterative improvement of augmented instruments.

Evolution of HASGS

Nilson describes in detail the process of design and development of digital instruments, dividing its development process into *Design Time* and *Play Time* (Nilsson 2011). *Design Time* is compared to composition, a process that takes place "out of time" in which design and implementation decisions are made. *Play Time* is when the instrument is played, allowing its evaluation in terms of the sensation of the performance, and the possibilities and expressive sense that the instrument can provide.

Playing an instrument is therefore an important part of its design process, with performance being the exploration of instrumental possibilities and how the instrument evolves in relation to what is intended, in terms of response and feedback. Waisvisz argues that it is essential to stop the development of an instrument, take a step back in the construction process and start playing as it is, composing and exploring its limitations (Waisvisz 1999).

The first HASGS prototype was made using an Arduino Nano plate attached to the instrument's body and mapping the data from a ribbon sensor, four-button keypad, a trigger button, and two pressure sensors. One of the pressure sensors was located on the saxophone's mouthpiece in order to detect the pressure exerted on the mouthpiece during a performance. The remaining sensors were positioned within reach of the left and right thumbs. This placement proved to be quite efficient because during a saxophonist's performance, these two fingers are quite free in relation to the keys, and their positioning contributes to the support and stabilisation of the instrument. The communication between the Arduino board and the computer was programmed through a serial port using a USB connection and by running a Node.js application simulating a MIDI port to receive data from the USB port and sending it to a virtual MIDI port.

A second prototype was proposed maintaining the characteristics previously described, and with the intention of incorporating an accelerometer into the augmented instrument to produce wider gestural capabilities and, above all, to record some kind of biofeedback with the use of a Myo device. The communication between this device and the host computer was carried out through the Bluetooth protocol supporting the mapping through the object [Myo] for Max/MSP (written by Jules Françoise). The Myo armband technology was used to collect data from an accelerometer, gyroscope, orientation of quaternions, and eight electromyograms. The analysis of Myo's behaviour made it possible to collect gestural data and observe performative characteristics specific to the different types of saxophones within the instrumental family. This proved to have an enormous potential for characterising involuntary gestures, as well as for collecting and embedding biological data of different works.

The third prototype of HASGS swapped the previous computing board for an ESP8266, allowing wireless communication between the augmented instrument and a host computer. The elements of the system were connected via an API and linked by a phone hotspot serving as router. Regarding the set of sensors, two knobs were added allowing the control of independent volumes, for example. During this process, and by optimising the use of Myo as an optional element in the augmented system, we achieved more stable results using the External Object for Max/MSP named Myo Mapper (developed by Balandino di Donato).

Several performance opportunities with new and older repertoire led us to include an ESP32 card, providing Bluetooth and Wi-Fi connectivity. For a better attachment to the instrument, we opted for a digital fabrication

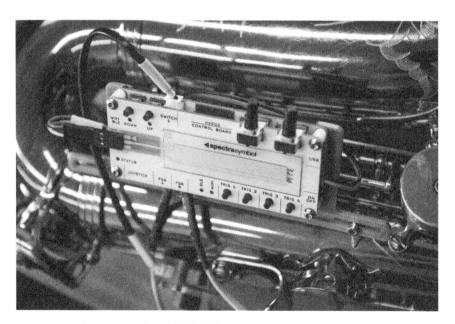

FIGURE 2.1 Current version of HASGS

Photograph by João Roldão.

solution that could be directly integrated into its body, taking advantage of the insert points available in the protection side plate. In terms of sensors, several improvements and updates were made in addition to those from the previous version, including up & down selectors, an axis joystick, a piezo sensor, a connection selector, an accelerometer and gyroscope, extra trigger switches; and some status LED indicators (for multiple functions).

The Myo Armband resource was abandoned due to the fact that the analysis of biofeedback data was not used as a resource by most of the composers that came into the project.

Mapping

Mapping constitutes the entire invisible part of the instrument, or the whole process from the physical gesture to the sound that is heard (DeLahunta 2010). In contrast to an acoustic instrument, an augmented instrument or

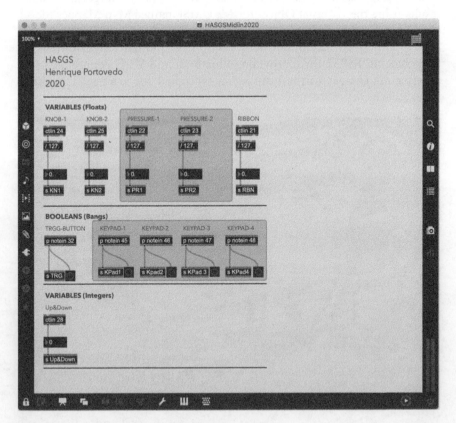

FIGURE 2.2 Data Mapping for HASGS

Photograph by the author.

electronic equivalent introduces an arbitrary factor into the design because the properties of its material and its shape do not determine the sounds that can be emitted. For this reason, the perform can experience a feeling of freedom in determining how a certain gesture creates or modulates a sound or timbre.

In the process of developing support and encouraging the creation of repertoire, a table of instructions was presented describing the possible communications between the sensors and the software. This was sent to several composers, suggesting a standard in relation to how the software could be used, giving preference to the programming in Max/MSP. Thus, the table indicated the objects and attributes related to the mapping of each sensor. An abstract in Max/MSP has been produced for this purpose.

Instrumental technique

Thinking about a general definition of instrumental technique on an Augmented Instrument, when relating it in comparison to an acoustic instrument might be considered very auspicious. The use of the augmented instrument's resources, at least during development, means that for each work or each composer there is the possibility of changing the data treatment of each sensor in a very flexible way, thus breaking general linearity. Because the influence of the mapping decisions is fundamental, it seems appropriate to mention that each work has its own instrumental technique and not just the instrument itself.

Writing code is a means to concretise and execute a composition and, when it is an integral part of composition, it creates and defines the rules of how the work will sound. It can also be necessary to analyse how a particular musical style or way of thinking about the musical work may be integrated into certain hardware. The following stablishes the principles and variety of possible instrumental techniques for each work.

Repertoire[3]

In addition to promoting contemporary notational developments, the creation of new repertoire suggests different approaches for using and extracting data from the existing sensors in the instrument based on flexible mapping potentiality. It is worth noting that the development of unconventional notation, due to the necessity of giving an indication of the action for the different sensors, is not dependent on the technology nor on the control of the devices associated with new instruments for the production of mixed music.

The notation of musical elements has continually evolved over time in line with the desire to produce new sounds or textures, a development that has also contributed to the increase in new instrumental virtuosities. When acoustic instruments are played in unconventional ways, the result can sometimes even sound like electronic music (Roads 1015). In relation to the new repertoire for augmented instruments, and, more precisely, with regard to this augmented saxophone system, it is necessary to underline the presence of multiple layers of information, something that is not common when writing for a monophonic instrument.

When analysing the compositional processes for instruments such as HASGS, it is important to evaluate the contributions of the different sound materials in isolation, as well as the unity of those materials in the final composition. This duality is raised by the fact that the different roles can be separated: as the composer has a concept, the programmer seeks to execute it in code. The code being executed and the composition interpreted by the machine and performer are cumulative steps leading to it being heard by the listener. However, it is possible that the composer and the programmer may collaborate to adjust the code and composition at the expense of the sound result. In this case, it is natural that the compositional concept can be adapted. The programming language itself, the complexity of building the algorithm, the programmer's decisions – all of these can be interpreted as compositional. As a result, the composition that ends up being shared, may raise the question of whether authorship can be unilaterally attributed, especially given that the role of a programmer may not be limited to engineering. In summary, the programming language used to achieve a given musical result is important, as is the programmer's personal ability (or vocabulary) and the dialect that they can speak within that language, and all can determine which musical ideas can be expressed.

The following works developed specifically for HASGS were composed between 2015 and 2020, some resulted from the presentation of the project at conferences such as EAW,[4] ICLI,[5] SMC,[6] and ICMC.[7] Of the works developed for the project, the four presented here demonstrate the typology and aesthetics of the repertoire based on the loop composition technique. "Indeciduous" by Stewart Engarts and "Disconnect" by Rodney Duplessis resulted from a period of residency at the University of California, Santa Barbara, from January to April 2018. "Cicadas Memories" by Nicolas Canot was written to be premiered at ICLI 2018 and was updated in 2020.

"Indeciduous"

"Indeciduous" is performed as a free blues on an electronic drum loop. Durations of different phrases are given as suggestions, as are musical gestures based on improvisational fluency. The sound pitches are performed in order to be part of the recorded loop and consequently triggered by the performer. The action of the looper is managed through the trigger button, suggesting a certain inactivity during the moments when the buffer of the looper returns the previously recorded material.

In terms of instrumental technique, the work makes use of HASGS in a very organic way integrated with the acoustic text, with no great complexity in terms of dynamics control. Above all, the fact that it is a semi-improvised work makes the management of action-inaction timings, also measured according the need to trigger commands or not.

This work does not consist of different sections based on presets; the mapping is constant through its structure. This allows us to assume that there is a linearity in terms of mapping, with the actions of the technical control of HASGS in constant relation to the behaviour of sound and effect along the work.

The augmented system allows decisions to be made about which elements or phrases can be recorded for the creation of loops. In this sense, the instrumental discourse determines the integral texture of the work, even if it is without colouring or electronic effects. The constant presence of a drone with regular pulse beats creates an atmosphere in which the discourse ends up being adapted, and, therefore, there are several sound layers: a constant one; one that introduces new material; and one that is an evocation of past material. This last layer is very dependent on HASGS, since the position of the

FIGURE 2.3 Notation Example of Indeciduous

Photograph of the Score by Stewart Engarts.

TABLE 2.1 Data Mapping of Indeciduous

Potentiometer 1	Gain of the Saxophone
Potentiometer 2	General Volume
Pressure 1 (Left Thumb)	Size of Looping Window
Pressure 2 (Right Thumb)	Loop Location within Looping Window
Ribbon	Reverberation Time (seconds)
Trigger	Start/Stop Recording of Loop
Keypad 1	Start Drum Machine
Keypad 2	Stop Drum Machine
Keypad 3	Adds and Trigger Events
Keypad 4	Stop All Loops

thumbs will control the reproduction of the buffer, and it is also possible that these past elements are presented in retrospection, creating a phenomenon of uncertainty in relation to the electronic material and its relationship with the instrumental material played. As previously mentioned, the notation of the work is referential and works as a suggestion of a harmonic field, giving a degree of performative freedom. All musical events are entirely controlled by the soloist.

"Disconnect"

"Disconnect" takes advantage of the discreet and continuous control provided by HASGS, in order to make the performance of electronic processing elements more organic. The electronic component consists of a set of buffers for recording and reproducing the saxophone material, including the loop of that material and a bank of filters. The latter is defined with formants for three different vowel sounds: ⟨ə⟩ ("uh"), ⟨ɪ⟩ ("ih"), and ⟨ɑ⟩ ("aw").

We can consider that there are two distinct sections in the work: a first part based on a loop of melodic material, which, with the accumulation of these same elements, creates a web of complex harmonic relationships, albeit in a language closer to tonal development; and a second part in which some of the material from the first part is exposed, but where the chain of relationships caused by the constant loops is mainly about timbre and where the range of formants is evident. In this second section, the sounds produced have a "windy" character, amplifying the wind inside the aerophone and the sounds of the consonants "sh", "t", and "f".

In terms of mapping, this work maintains the linearity between the controller and its functionality throughout. Note that pressure sensors were not used for any parameter of control in this work.

In performative terms, "Disconnect" applies several parameters of electronic effects over the saxophone sound, not only through formants, as

FIGURE 2.4 Notation Example of Disconnect

Photograph of the Score by Rodney Duplessis.

TABLE 2.2 Data Mapping of Disconnect

Potentiometer 1	Gain of the Saxophone
Potentiometer 2	Gain of the Looper Playback
Pressure 1 (Left Thumb)	
Pressure 2 (Right Thumb)	
Ribbon	Playhead Velocity
Trigger	Start/Stop Recording of Loop and Start Playback
Keypad 1	Filterbank 1
Keypad 2	Filterbank 2
Keypad 3	Filterbank 3
Keypad 4	Stop All

previously mentioned, focusing on aeolian sounds, but also by creating variable layers of loops and exploring the attack of transients with different articulations. The sound of the acoustic instrument is the source for the elaboration of the electronics, controlled by the augmented system and generating several layers, sometimes over actual time, sometimes over past events. In terms of notation, we have exactly the same characteristics of the work previously analysed, with the sharing of both a traditional and an expressive notation inherent to the augmented system. All the action of the performance unfolds under the absolute control of the performer.

"Cicadas Memories"

Composed by Nicolas Canot, "Cicadas Memories" is much more an improvisation process than a written piece of music. The work explores a method that introduces performatively unusual ways of thinking about music in which live music is controlled and altered by updating the past. This means that the performer's gesture will – after a one-minute delay – change the texture of the current electronic sounds providing a sonic background to the melodic discourse and rhythmic impulses of the saxophone. Therefore, the performer needs to develop two ways of thinking simultaneously during the performance: the first refers to the present (i.e., the standards imposed by the software but created by the past action of the performer), and the second refers to the future (i.e., gestural connection with the sensors). Thus, the performer has to deal with two temporalities, generally separated during the act of performing live music, by both determining the future score and improvising on past gestures in the present time. "Cicadas Memories" can be defined as a multitemporal feedback loop. With regard to the sound and musical context, multitemporal feedback explores the player's thinking as a process, perhaps influenced by Di Scipio's thoughts (Di Scipio 2015), instead of "written music" movements. To give the performer sufficient freedom, the design of the interaction between sound and gesture in the HASGS is generally not as deterministic as it is in acoustic music performances associated with new instruments for the production of mixed music.

The fact that "Cicadas Memories" is composed for an augmented instrument is important because it emphasises the relationship between the type of values produced by the sensors:

1) Modulating Variables VS Boolean Values;
2) Continuous Stream of Data VS Fixed Values;
3) Freedom of Performer's Body Gestures VS Necessity to Interact with sensors of the fingers.

This means that the performer's gestural activity through the sensors determines the way the instrument is performed. The additional performance of the sensors in the instrument body modifies or alters performance patterns, making it evident, within the scope of the composition, that the four buttons on the keypad could be thought of as a 4-bit data flow generator. Because 4 bits means 16 different values ranging from 0 to 15, it quickly became clear that these 16 values could be retrospectively related to the sixteenth note of sixteenth notes in a 4/4 measure, given the traditional structure of Western music.

The sonority of the electronics included the creation of sounds from nature, for example bizarre creatures, various sounds of foliage, cicadas, the splitting of wood, and glissandos with bird sounds. This spectrum intends to create a kind of soundscape relating to the soloist's performance decisions.

The values taken from the sensors were normalised between 0 and 1 and the data presented is constant throughout the work, with no differences between classes of variable mappings on the timeline, but only when a preset is selected. The abbreviation nm stands for normalised.

[p +delay] synth:

(pr1nm/kn2nm): delay time;
(pr2nm/rbn2nm): delay feedback;
(kn1nm/kn2nm): delay resonance;

(pr2nm/kn1nm): overdrive 1 gain;
(pr1nm/kn2nm): overdrive 2 gain;
(pr1nm/kn2nm): synth output gain;

[p all-sqnzr] synth:

kn1nm: synth output gain;
kn1nm: right channel delay in samples (stereo width);

NV1: connected to KP1 inside the [p distrib] sub-patch, it increments the tab note-value to adjust the allpass filters time (note values converted to ms) each time the *binary* combination of the Keypad 1 is equal to 0 or 8;

NV2: Keypad 2 binary combination equal to 1 or 4;

NV3: Keypad 3 binary combination equal to 2;

NV4: Keypad 3 binary combination equal to 4;

S1 to S16 activates each step of the sequencer via the keypads (4 steps/sixteenth notes for each PAD in relationship with the display in the main patch);

TRG resets all sequencer's steps to 0;

[r seq_step] adjusts the number of steps (sixteenth notes, from 1 to 16) of the sequencer in relationship with the *binary* combinations (inside the [p distrib] sub-patch). This function might appear complex and requires some time using the keypads only:

KP1 has a value equal to 8;
KP2 has a value equal to 4;
KP3 has a value equal to 2;
KP4 has a value equal to 1;

The different *binary* combinations of the keypads values can produce every possible loop length from 1/16 to 16/16. Of course, only the steps (orange squares are active steps) included in the loop length will be played;

[p glitch-synth] synth:

cnt1 to cnt16 (in relationship with the *binary* combinations of the keypads) control some synced frequencies defining the gain of the incoming signals in the filters as well as the two samples length, start and end points, speed/pitch in regard to the tempo so, in sync with [p all-synth] and [p rain-osc] patches;

KP1 sets the centre frequency of the resonant filters in a random way;

pr1nm sets the output gain for each sampler;

kn1nm adds some kind of saturation to the signal (left sampler);

kn2nm adds some kind of saturation to the signal (right sampler);

[p rain-osc] synth:

(pr1nm/pr2nm): synth output gain;

kn1nm: range of the random starting frequency (left) of the glissando;

kn2nm: range of the random starting frequency (right) of the glissando;

pr1nm: added value to the starting frequency (left) of the glissando;

pr2nm: added value to the starting frequency (right) of the glissando;

rbn1nm: added value to define the ending frequency of both glissandi (left and right have different values even if they share the same controller);

kn1nm: attack filtering/smoothing (left);
kn2nm: attack filtering/smoothing (right);
(pr1nm/pr2nm): allpass filters gain.

"Verisimilitude"

Composed by Tiago Ângelo, "Verisimilitude", written for tenor saxophone and HASGS system, uses a single loudspeaker placed in front of the performer at the same height as the saxophone bell. A mixture of acoustic and electronic sound sources (processing and generation) based on computer music techniques are conducted in three sections: A, B, and C. Each of these sections is presented with its own specific processors and generators through the implementation of different mappings at the control level, either from the HASGS controller, or from real-time sound analysis.

Starting from an expressive notation, we opted for a pitch suggestion that establishes a kind of drone in the text reserved for the soloist part. This note must maintain the continuous sound with slight oscillations of a quarter tone, as indicated in the score. Oppositional to this is the final section, where the oscillation and permutations of sounds performed must conform to what is noted in the score with just-tuned notes. The sound continuum, composed of long notes that are almost drones, is dynamised by the oscillation of pitches of tone from that same note. Additions to these variations of timbre, through the processes controlled by the augmented system, are presented graphically in a continuous way and changing between: reverb, spectral delay, detuned spectral freeze, voice formant resonators, amp modulation, spectral pitch shift, and distortion.

Given the number of acts of manipulating the HASGS sensors, it was decided to take the transposed note G as the fundamental of the work, from which the oscillations of timbre and tuning take place.

In "Verisimilitude" electronic sound events arise from the sound emitted by the soloist instrument itself. This characteristic of colouring the timbre, combined with small oscillations from the fundamental note of the entire work, creates fairly rich environments. The control of the electronic parameters is

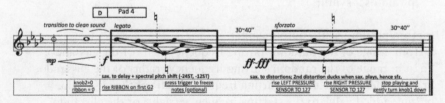

FIGURE 2.5 Notation Example of Verisimilitude

Photograph of the Score by Tiago Ângelo.

entirely carried out by the augmented system with no place for the automatic generation of effects from the computational algorithms apart from a looping effect auto generated in the last bars of the piece. The number of parameters that can be manipulated, together with the sound elements from the feedback created from the movements in front of the amplification column, generate a large number of electroacoustic effects.

"Comprovisação n°9"

Comprovisador is a system designed and programmed by Pedro Louzeiro that allows an interactive mediation between soloist and ensemble using machine listening, algorithmic composition and dynamic notation in a networked environment. As the soloist improvises, the *Comprovisador* algorithms generate real-time notation that is immediately read by a group of instrumentalists, creating a coordinated response to the improvisation. The interaction is mediated by a performance director, in this case the composer himself, through algorithmic parameters. The implementation of this system requires a series of computers connected in a network in order to display the notation in separate parts for each of the ensemble's players.

In "Comprovisação n°9", HASGS is used as a musical interface with a dual action: Feed *Comprovisador*'s algorithms with improvised musical material via an acoustic instrument; Control several parameters through its sensors and controllers, adopting some of the performance director's mediation tasks to the benefit of the interactive flow.

The synergies between HASGS and *Comprovisador* allow a degree of greater interactivity, between the improviser and the ensemble musician sight-readers, that is to say, between the improvised material and the composition in real time. By enabling the soloist to control selected parameters of an expressive, compositional and formal nature, a consequent interaction is expected. In addition, it allows the performance director to be more aware of the macrostructure, than when controlling all parameters.

A performative plan outline is obtained through the predefined algorithmic parameters and consequent control mapping. Each preset produces different types of musical response, ranging from the reaction of synchronised rhythmic clusters between the ensemble members, to intricate micro-polyphonic textures. The keyboard of the HASGS (keypad) allows the soloist to navigate through the presets of *Comprovisador* according to the performative plan plotted, while maintaining the relationship with the thematic material stripped at the time, and some of the other controllers, such as the ribbon, trigger and pressure sensors, allow them to control parameters of dynamics, harmonic density and instrumentation, and height and pulse recording, among others. In addition to that previously described, HASGS is used for the ability to trigger actions of algorithmic transformation, including the capture of melodic contours or evocation of passages and previous musical material, the latter being passages previously played or pre-composed.

The performance aims to create a context where the composed and improvised elements coexist in an aesthetically coherent interdependence, taking advantage of the possible synergies between a composition and notation system in real time and a hybrid acoustic instrument, enabling an increase in the level of interactivity. The flow of the interaction is complemented by the soloist's reaction to the composition's response, and intensified by the presence of a performance mediator, establishing a complex dialectical relationship. The fact that HASGS, in this case, is used in an environment different to that originally expected – the execution of works that presuppose control of electronic sound parameters in real time – presents challenges and creates opportunities from the experience of the soloist, since the interaction, as previously described, is somewhat more instantaneous than dealing only with composition and notation algorithms in real time.

Conclusions

The HASGS system was developed in close collaboration with several composers and evolved according to compositional ideas developed during different stages of the project. Whilst analysing the panoply of augmented instruments that emerged in the NIME context, several problems were identified at different levels.

So far, the different stages of the development of HASGS have suggested the addition and subtraction of technological resources at the expense of its use as compositional and performative elements. Although the initial desire was to build a system that would emulate an Electronic Wind Instrument, I realised after initial testing that the evolutionary path would have to take another direction. In this sense, "Reduced Augmentation" is advanced as a means of managing and balancing the use of technology. The main concern designing the hardware system was making it unobtrusive to the acoustic instrument, working instead as an additive kit or set.

We tried to stimulate free composition, although the electronic resources and sensors available always include certain limitations. Technological possibilities were added, while others were removed based on the elaboration of the pieces, suggestions from composers, peers, and stages of performative practice. The definition of an instrumental technique largely underlies the aesthetics of the pieces that constitute the repertoire of an instrument. The repertoire developed for HASGS is an example of the creative variety that mapping supports. Consequently, the difficulty of accurately defining a standardised instrumental technique is enormous, even when the relationship between an augmented system and an acoustic instrument allows us to establish similarities, insofar as they show how composers made similar use of the technology.

Particularly in the context of this book, it shows how live looping as a compositional technique is integrated into the selected works. It also analyses how live looping as an aesthetic profile relates to Electronic and Mixed Music.

Acknowledgments

HASGS research was supported by National Funds through FCT (Foundation for Science and Technology) under the project SFRH/ BD/99388/2013, from 2014 to 2019. Fulbright has been associated with this project supporting the research residency at the University of California, Santa Barbara. I acknowledge all the composers mentioned here, and my actual research centre INET-md for the support on continuing the project these days.

Notes

1 Parts of this chapter were submitted to NIME2021 (New Interfaces of Musical Expression Conference) as a historical overview of HASGS.
2 Institute for Research and Coordination in Acoustics and Music.
3 All works presented here can be further explored and heard at https://www.henriqueportovedo.com/hasgs/.
4 Electroacoustic Winds Conference.
5 International Conference on Live Interfaces.
6 Sound and Music Computing Conference.
7 International Computer Music Conference.

References

Bowers, John, & Philip Archer. "Not hyper, not meta, not cyber but infra-instruments". Paper presented at the Proceedings of the 2005 Conference on New Interfaces for Musical Expression, Vancouver, Canada.
Burtner, Matthew. "The Metasaxophone: concept, implementation, and mapping strategies for a new computer music instrument". *Organised sound* 7, no. 2 (2002): 201–213. doi:10.1017/S1355771802002108.
Collins, Nick, Schedel Margaret, & Scott Wilson. *Electronic Music* (Cambridge: Cambridge University Press, 2013).
DeLahunta, Scott. "Shifting Interfaces: art research at the intersections of live performance and technology" (Doctoral thesis, University of Plymouth, 2010). Retrieved from http://hdl.handle.net/10026.1/2711.
Di Scipio, Agostino. "The Politics of Sound and the Biopolitics of Music: Weaving together sound-making, irreducible listening, and the physical and cultural environment". *Organised sound* 20, no. 3, (2015): 278–289. doi:10.1017/S1355771815000205.
Impett, Jonathan. "A Meta Trumpet(er)". Paper presented at the ICMC, Ann Arbor, 1994.
Kimura, Marl, Rasamimanana, Nicolas, Bevilacqua, Frédéric, Zamborlin, Bruno, Schnell, Norbert, & Emmanuel Fléty. "Extracting Human Expression for Interactive Composition with the Augmented Violin". Paper presented at the International Conference on New Interfaces for Musical Expression (NIME 2012), NA, France.

Machover, Tod. "Hyperinstrument: Musically Intelligent and Interactive Performance and Creativity Systems". Paper presented at the ICMC, 1989.

McPherson, Andrew. "The Magnetic Resonator Piano: Electronic Augmentation of an Acoustic Grand Piano". *Journal of New Music Research* 39, no. 3 (2010): 189–202. doi:10.1080/09298211003695587.

Miranda, Eduardo R., & Marcelo Wanderley. *New Digital Musical Instruments: Control and Interaction Beyond the Keyboard* (Computer Music and Digital Audio Series: A-R Editions, Inc., 2006).

Neill, Ben, & Bill Jones. "Posthorn". Paper presented at the 28th International Conference on Human Factors in Computing Systems, CHI, Atlanta, Georgia, USA, 2010.

Nilsson, Per A. "A field of possibilities: Designing and playing digital musical instruments" (Doctoral thesis, Gotemborgs Universitet, 2011).

Normark, C. Jorgen, Peter Parnes, Roberto, Ek, & Harald Andersson. "The extended clarinet". Paper presented at the International Conference on New Interfaces for Musical Expression: 11–15 July 2016.

Palacio-Quintin, Cléo. "The hyper-flute". Paper presented at the NIME, McGill University Montreal, Canada, 2003.

Penny, Jean. "The Extended Flautist: Techniques, technologies and performer perceptions" (Doctor of Musical Arts, University of Melbourne, 2009). Retrieved from https://www.jeanpenny.com/uploads/5/5/4/3/55434199/penny_the_extended_flautist.pdf.

Roads, Curtis. *Composing Electronic Music: A New Aesthetic* (NY: Oxford University Press, 2015).

Schiesser, Sébastian, and Jan C. Schacher. "SABRe: The Augmented Bass Clarinet". Paper presented at the NIME, 2012.

Thibodeau, Joseph, & Marcelo Wanderley. "Trumpet Augmentation and Technological Symbiosis". *Computer Music Journal* 37, no. 3 (Fall 2013): 12–25.

Waisvisz, Michel. "Gestural round table". *STEIM Writings* (1999). Retrieved from http://steim.org/media/papers/Gestural%20round%20table%20%20Michel%20Waisvisz. Pdf.

3
INTERACTION AND REACTION

Reflections about performance, composing, and live looping

Iury Matias de Sousa

Live looping and the artist

Live looping in the context of a performance is for me a resource that has contributed to developing other possibilities for my creative work, becoming a musical partner and, later, a compositional partner. I use both instrumental music and songs for solo and group performances, creating soundscapes, "musical environments", arrangements and compositions, either previously constructed or, sometimes, improvised live. I understand the audience's reactions during the performance to be a way of communicating with the music that is being performed at that moment, and with how they are feeling about it, so the soundscapes are also created from these reactions, from my interaction with the public.

The use of live looping came, among other things, from a need to listen to my compositions, isolate some elements and create others some elements and creating others, from harmonies, melodies, to rhythms built note by note, in an experiment that continues to this day, to this day, the attempt to bring memories, sensations, feelings, places, noises, etc., together, interacting with each other and with me, and to communicate, through performance, in a dialogue with the audience. This process has re-signified itself to me over and over again, making the use of live looping an essential device for my work.

My concerts using live looping have been held in various contexts, from bars and restaurants, to cultural associations, concert halls, theatres and large stages, and, due to the Coronavirus (COVID-19) pandemic, via live streaming performance, and some recordings of these performances are published on YouTube. Usually, performances of my own songs and instrumental compositions can be carried out in spaces of any nature, as long as the necessary

DOI: 10.4324/9781003154082-5

sound equipment is available, with one or other change regarding the audio technique in order to better execute the programme. My inspiration comes from Brazilian music in its different aspects, especially its indigenous roots and connection with African culture, in line with the influences of jazz.

I understand that my artistic work has been tracing the path "from front to back", as I first studied music and absorbed knowledge about theory, harmony and instrumental technique, as well as music history, especially Brazilian Popular Music and Jazz, and then, many years later, I began to investigate the music that affected me inside. That is where I discovered that the music that I was creating was directly connected to elements coming from the rhythms of African drums, amid simple melodies, tonal harmonies and more, which I could see very vividly were present in the popular music of northeast Brazil, until more recently, where I have found myself strongly connected with "the sounds of the forest" and the "indigenous vein" that pulsates in my music.

Record, play, and stop

The aim of this chapter is to share the phases of my creative process, my performance, and my way of using live looping, as well as describing my relationship with this technology, going through the "turning point" where technology meets and is involved in composition, until reaching the latest phase, which consists of the recording of some of work. It is the objective of this chapter to understand, in addition to these processes and relationships, the way in which the compositions were organised and what concessions were made in relation to live looping, so that I could get the songs to the point of recording.

The methodology used for the construction of this chapter was based on the analysis of my own performances, essays, compositions and notes, in parallel with the critical appreciation of other artists and the public present at the performances. The analysis of audiovisual material was considered, highlighting the musical performance, the use of live looping, the observation of what was planned in relation to what was performed in concerts, and the discussions with and opinions of engineers, audio technicians, composers, and performers who use live looping.

In this chapter I will talk about my instruments and the different set-ups used from the beginning of the performance's creation up until now, as well as the difficulties faced, both in the audio-technical scope, regarding the sound, and in the performative scope. I will also examine relationships with audio professionals and interaction with the public, and describe the entire course of the compositions, from the performance using live looping, through rehearsals with other musicians, and to the moment of recording them in the studio.

The analysis of three compositions of my own will be discussed: *Inês e a Ilha* (2014), *Akiré* (2015), and *Daqui a Você* (2015), and also arrangements of works by other authors (not specifically mentioned but in general), in order to reflect on the process of building arrangements using live looping and the way in which this technology was involved in the development of some compositions, as well as trying to understand the differences between its use for each case.

Instrumentation and setup

The equipment used was changed through the course of experiments and the construction of the performance with live looping, basically keeping just the guitar, which is electroacoustic, and the electric guitar. The first loop pedal was a model with two footswitches, with a mic channel and an instrument channel, which recorded on separate tracks and had P10 and XLR outputs. This pedal was replaced by another that I still use today, which has four footswitches, with a microphone channel, an electric guitar channel and a P10 mono auxiliary channel for the instrument, and three outputs: two P10 mono (L and R) and one P10 stereo.

The electric guitar setup features a wah pedal, a compressor, a drive pedal, an octave pedal, a chorus and tremolo pedal with two footswitches and independent controls for each effect, a delay unit and a reverb. I use a pedal for voice that is also a D.I. – that is "Direct Input" – electronic device typically used to connect a high-output impedance, line level, unbalanced output signal to a low-impedance – for guitar, which has a loop, another octave pedal that I use along with a sound retainer pedal, and an effects processor and digital amplifier simulator for guitar, which also has a loop. I have two cardioid and unidirectional microphones, one with an "on-off" switch, which I use more often, and another without this switch, which I used at the beginning of the performance-building process.

I first saw a bass player using a loop feature present in a delay pedal on a television show in 2000, and from there I became interested in the technology. In 2003 I purchased an electric guitar effects processor that offered a 7.5-second loop, which, despite the poor quality of both this feature and the sound of the equipment in general, was an interesting experience. In 2006 I managed to buy another guitar processor with an 8-second loop, this time of very good quality but very limited in terms of the loop, which made me lose interest in the subject for a while. It was only in 2012 that I bought a loop pedal to use as a resource for works with the singers and soloists with whom I worked at the time, mostly for recording and improvising over live bases.

Later, I saw a performance by the Cameroonian bassist Richard Bona where he used live looping during his concert to create overlapping voices.

His relaxed and musical way of manipulating the voice to create the musical context into which he inserted this overlay caught my attention, and, although I was not using the vocal overlay feature, this performance was for me a reference for the creative possibilities of the loop. Subsequently, when I was already using live looping in my performances, I watched a performance by the African guitarist Lionel Loueke, with whom I quickly identified due to the fact that he simulates percussion instruments and other elements present in nature using sounds produced with the mouth, together with the overlays of melodies and chords that he plays on guitar. I also came to know the work of the Brazilian multi-instrumentalist Munir Hossn, who also makes use of live looping in his performances with guitar, bass and voice as an instrument, playing songs and melodies to the sounds of African drums present in Brazilian Afro-religious culture.

Videos of complete concerts of mine, and also of Modulatus, a duo project with Portuguese singer Laura Rui, were analysed, including both live performances and live-streaming format available on YouTube, as well as short videos available on my own social networks and on your channel. Video, audio and notes were also analysed, the content of which was built over the previous four years, including live looping experiments and the development of arrangements and compositions using this technology, which form part of my personal collection, not available to the public.

Listening to and observing this material made it possible to establish both the timeline for the development of skills acquired through the use of live looping, and the understanding of the process triggered by the influence of this resource on performance.

Overdubs, technical difficulties, and solutions found

When I finally got a live looping device, my setup consisted of just a guitar and a microphone, leading me to use loop features such as reverb, delay, chorus, etc. It was a pedal that offered two footswitches, which made the relationship between playing, recording, changing recording tracks, pausing the loop, playing again, and erasing a track, among other functions, very difficult. It was then that I bought equipment from another brand that had four footswitches and a channel dedicated to the electric guitar, with amplifiers, effects and other individual resources, as well as individually adjustable microphone resources, which changed the perspective of use and, consequently, the performance.

This loop was developed for the electric guitar – although it works perfectly with the guitar, which I often do – so I soon started experimenting with my analogue pedals setup for electric guitar connected to the loop, getting a schematic like this:

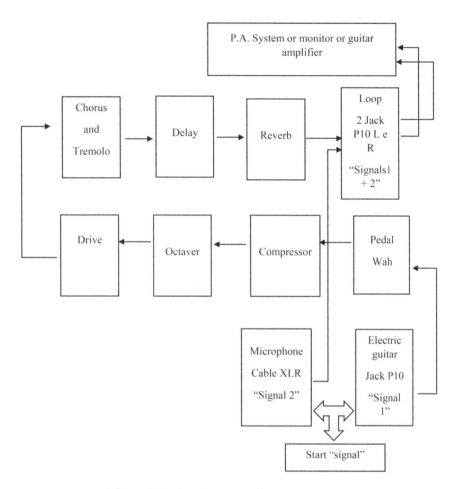

FIGURE 3.1 Signal flow of the first loop and effect pedals setup

From that moment on, a period of much experimentation began, both in terms of creation and the exploration of the possibilities of all the equipment, but mainly in the search for an acceptable signal quality, since the first consequence when connecting my guitar setup to the loop was an amplification of noise added to a loss of definition of the instrument's timbre. Initially there was no way of knowing for sure what was happening, because in addition to my having limited knowledge of the audio technique, there was no financial resource with which to acquire the proper equipment for powering and connecting the pedals and sound amplifier. The compressor pedal was developed by a friend who built it to "improve" the sound of the instrument by enhancing some frequencies, which helped to "disguise" the noise, giving more gain to the guitar signal. The combination of this pedal with the drive also gave an interesting result, in addition to helping to reduce signal loss, hence the choice to place it at the beginning of the pedal order.

The replacement of less expensive parts in the setup was one of the strategies adopted, for example the connection cables with gold plugs at the ends and more robust wires that provided an improvement in the signal flow, leading to a decrease in noise which, although it did not completely solve the problem, contributed positively to the final sound result. Another alternative was to modify the position of the pedals in the chain, which was surprising due to the discrepancy in the result of the timbres in the combinations of connections between the pedals, drastically altering the sound of each pedal. Most of the attempts gave frustrating results, especially when the modification went through the reverb and delay pedals, as they were responsible for "tempo" effects, but when combined with the other pedals, the timbre of the electric guitar itself was compromised.

Another pedal that presented a delicate situation in relation to its position in the chain was the octave, as the first model used in the pedal chain allowed eleven combinations of the pure sound of the instrument, octaves below and octaves above, making a combination work for some connections and not others. With the analogue gear, it was impractical to maintain a pedal setup that required constant manipulation during performance, so the octaver had to maintain its position in the chain, just like the reverb and delay. There was the possibility of changing the position of the chorus and tremolo pedal, a model with two footswitches and separate parameter controls to independently configure and trigger each effect, which, despite making it quite versatile, also caused compatibility problems in combinations with other pedals to surface in the signal result.

It then remained to try changing the position of the wah, compressor, and drive pedals, which were exactly the pedals that would preferably not come out of their positions because they change the signal gain, causing the instrument's timbre and all the pedals to undergo considerable changes. After trying different positions to alleviate concerns, it became clear to me that the first pedal in the chain should be the wah, followed by the compressor, the octave, and then the drive. Thus, the guitar timbre would be maintained but receiving more gain, both from the wah pedal and from the compressor, causing the octave to send the signal, modifying the octaves before reaching the drive. Therefore, the signal went to the modulation effects already properly balanced to reach the loop.

The loop configuration that I used in the performances with this equipment arrangement is organised in two ways: the electric guitar parameters and the parameters for the voice. As this loop unit was developed for guitar and has an independent channel for the microphone, there are several simulations of amplifiers and effects, which allowed a "pre-amp" configuration of the guitar signal, making it possible to send the signal directly to the soundboard of a P.A. system in live performances. This equipment has two mono outputs (L and R) and one stereo output for the amplifier, allowing the signal to be sent simultaneously to a third distribution, which I usually use when there is feedback monitor equipment to have as an individual reference.

One of the biggest difficulties encountered, after the problems with the quality of the signal (which was never definitively resolved), was the transportation of the equipment, from venues in the city where I live to other cities, but mainly in situations that included air travel, because the weight (11.7kg) and dimensions (70–85 x 800 x 390 mm) of the gear are a major drawback. Faced with this complication, it became tempting to start using the guitar resources existing in the loop, so a phase of experimenting with timbres began, based on the effects and parameters that the equipment offered. This quest consisted of attaining the delay and tremolo times, reverb ambience, chorus timbres, drive and, since there was no octaver, compressor and wah pedal, I took advantage of the phaser and flanger present in the on the pedal loop to create and use other timbres in the performance.

When the electric guitar pedals left the setup, the loop became the only equipment on the floor, increasing its potential to be explored as far as my work programme allowed, initially being used in live performances to enable the creation of bases for improvisation, in situations when the guitar was the only harmonic instrument in the formation, especially in performances for voice and guitar. And, later, adapting my own compositions to this reduction of equipment, by starting to use the acoustic guitar instead of the electric guitar in many cases, which change position the loop once again to another place, where the resources would be limited in order to keep the timbre of the guitar as little altered as possible. At this point I started to remove any amplifier simulation, using the pure guitar signal amplified by its own preamp, adding only the reverb on loop pedal, and explored possibilities of the instrument's own timbre to create rhythms and noises that fed the soundscape of performance.

With the decrease in equipment and the use of the guitar, the possibilities with the loop were expanded, influencing both the creation of new arrangements with live looping and the conception of new compositions, and during this process I had the opportunity to acquire two pedals that have become – together with the loop pedal – the setup I have used the most to date, due to both their practicality and the scope of possibilities. These pedals are an octave, a brand and model different from the one I used before, and a sound retainer pedal, which work for me as "the bass" and the "keyboard (pad)" respectively, resulting in the following signal setup:

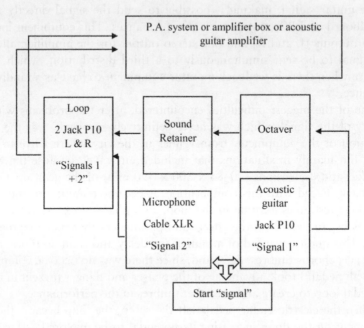

FIGURE 3.2 Signal flow diagram of the second loop and effect pedals setup

This setup became something of a standard for my performances, as after four years of constantly working with live looping, I had become familiar with creating improvised loops and had started to adapt the equipment to the situations. The new octaver keeps the first three strings of the guitar – or the electric guitar – with the natural timbre, changing only the three lower strings to acquire the timbre of a bass guitar, modifying the frequencies in such a way so that, depending on the equalisation, it can sound like an electric bass or even an acoustic bass guitar. The sound retainer allows control of the sustain of a note, chord or any sound extracted from the instrument in three durations: "short", "infinite" (from the "on-off" control) or "controllable", via the footswitch.

One example of practical use was, where previously I had used live looping to create bases and improvise during the songs, it started to be useful for building the instrumental layout of one of the parts of the song – with the harmony, the bass line, the sounds of a drum or percussion made with the mouth – so that the end result of the loop made it sound like a group playing. But as a music producer I came to realise that the crisp, purposefully constructed creation of loops didn't "had an impact" me, didn't interest me, didn't thrill me, because it sounded obvious or because there was no "element of surprise", and watching performances by musicians who used live looping in the same way confirmed this feeling and I realised that it didn't "affect" the audience either.

However, accompanying a certain singer, I was surprised by the sensation that came over me and by the reaction of the audience, when I realised that while I was going through the chorus of one of the songs in the repertoire, I recorded the chords on the loop pedal, saved it and continued playing the song. In the chorus, I recorded the bass line, and hit play again with the full loop at the end of the song. While the loop was playing I recorded artificial drum sounds produced with my mouth and soloed with my acoustic guitar over the chords, culminating in an artificial "fade out", made by manipulating the equipment's volume knob.

Between the professional need to have several different setups for each situation and keep the use of live looping technology in my favour whenever possible, and the fact that this technology started to be present in a lot of equipment for musicians, from instrumentalists and singers to DJs, I learned to extract loops according to what each setup offered. Due to the demand for solo concerts with a repertoire predominantly made up of songs, I started to constantly use equipment with the specific purpose of bringing a set of effects for voice, and a few effects for guitar, including the feature of creating loops which, although not recording on separate tracks, records up to 60 seconds with no limit on overdubs. With this voice equipment I connected my new setup in the same way as the previous one:

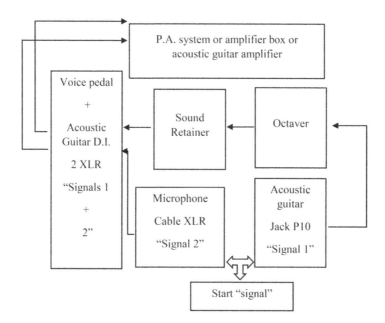

FIGURE 3.3 Signal flow diagram of the third loop and effect pedals setup

This setup has been the most used for my own solo concerts and situations in bars, concert halls and the like, for the last two years. The microphone was one of the elements of my equipment that never brought any problems, having been the same until mid-2018, when I started using a model that was comparable but with a very simple and very precious feature, which was the "on-off" switch. This simple feature helped to avoid feedback generated by the acoustics of the guitar which, although low – both in terms of frequency and intensity – occurred with the loop used in the previous set, and also worked with the live looping performance of the voice pedal, which is much more sensitive to feedback.

As this pedal is also a D.I. for guitar, even at low gain levels it causes feedback from my instrument, which is electroacoustic, making the whole sound more "delicate" in terms of microphonics, and for this reason the new microphone with its function switch makes it possible to reduce feedback. Also regarding the "on-off" switch, I found that using it to turn off the microphone during guitar loop recording was a strategy to avoid, as, in addition to feedback, there was the recording of an unwanted loop, which compensates for the absence of separate track recording feature on the loop pedal. With no financial means with which to acquire an all-electric solid body guitar, it is a big challenge to work with live looping in this equipment configuration.

As a guitarist, after being hired during this period to play and record the electric guitar in some works – although in lesser demand than as a singer-songwriter – the need arose to obtain a more practical use and more dynamic equipment that had the live looping feature, which is why I invested in an effects processor and digital amplifier simulator. This equipment does not have a microphone input and does not meet the loop demand for my performance, but it has been tested in order to become part of the setup, as it has a very comprehensive range of guitar effects and amplifier simulators, and works with Impulse Response technology, which allows you to create or copy sounds from the highest quality equipment available on the market.

The loop of this processor has a maximum duration of more than an hour, allowing infinite overdubs and signal flow configuration, which makes it possible to choose the sound of the loop to be turned on at the beginning of the signal flow, between the simulated pedals and amplifiers, or at the end of the chain, in addition to many other resources, having the signal quality as a great advantage, even with many effects used simultaneously. Its high-quality standard also allows the sending of the signal directly to the sound system table, without detriment to the quality of the instrument's timbre or even the effects, eliminating the need for a D.I. or guitar amplifier.

Interaction and reaction

The concert entitled *Cenário do Porto* was prepared in 2016, and in the years that followed, was presented in five cities in Portugal, also touring through

Cape Verde and Brazil, until 2020, when it was presented via live streaming due to the COVID-19 pandemic. *Cenário do Porto* consists of songs and instrumental compositions, in which the live looping technology is present and directly connected, as it was introduced in the work that already existed, becoming part of some arrangements, later changing some compositions and, later on, came to serve as a framework for other compositions.

In the songs, the loop assumed the function of "elaborating the soundscape", that is, of building an ensemble musical environment with only one musician, from one of the parts of the song, a musical excerpt or moment of improvisation. For instrumental compositions, the process was different, with loops being elaborated that became part of the composition and some of these loops later became the structure of the music itself. In this sense, concerns with sensors and pedals became a focus of the preparation of the performance, maintaining the characteristics of each composition and each arrangement, but preventing the triggering of loops and controlling of pedals from harming the spontaneity in the improvisatory character of the work.

Curiously, there is a concern that goes through the work of engineers or sound technicians, who generally have concepts, structures, and standards with which they prefer to work, often not being open to receiving a proposal with parameters that are not in accordance with their standards. This interfered time and again with the performance, becoming a constant concern, as there were situations in which the technician changed the volume of the signal sent to the mixer simply because the guitar could not have the bass sound generated by the octave pedal, because it was deemed as not being "natural". Similarly, I have had the sound completely cut off because the professional responsible did not like the frequency amplified by the sound retainer pedal and thought that, with that sustain, what was happening was a big mistake and would compromise the concert.

The most intriguing thing is that the problems caused by audio professionals because of the sound retainer pedal already passed an imposition of opinion, where I was confronted with the comment, "Look, I cut your sound! There was a horrible frequency here being fed back and it was very bad!" The concern with the reaction of audio professionals led, over time, to understand that it really is necessary not only to explain to them how the equipment works, but also, more importantly, how I use it, in order to bring these colleagues closer to the performance context and avoid unpleasant surprises for both parties.

Considering the detailed planning, the objectives outlined at the beginning of the construction of the performance, and its execution, the results analysed from the audiovisual material show that there was an evolution in the technical domain of creating with live looping, however the use of the resources of each piece of loop equipment was decreased following its acquisition. Observing the videos and audios from my own recordings, I believe that once familiarised with the equipment, and having gained from it the tools to create the loops defined for each composition, I chose to continue at that level of knowledge of the equipment, in order to ensure the quality of the performance.

It was interesting to notice that the improvisatory character of the performance took on a model for the compositions, becoming almost a standard, which did not happen with the arrangements for works by other composers, where improvisation is predefined, but not established. For example, in a performance of works that I have composed myself, the improvisation is defined –as to whether or not there will be improvisation, in which part of the music and how it will be done – subject only to changing in its duration. In an arrangement of another composer's work, improvisation can occur even without having been predefined, regardless of the arrangement, and disregarding the fact of having prepared, or not, an improvisation at some point for such music. Both in the improvisation and in the arrangement with the use of live looping, the compositions of others are "open", and it was possible to verify, from the analysed material, that many songs of this repertoire were completely different from the rehearsals, with many of them having an arrangement conceived through improvisation using live looping, consolidating itself in this way and starting to be executed always with the same conception or as similar as possible to it.

The first experimental recording made available on YouTube dates from 2016 and was of the song *Akiré*, composed in 2015 and arranged for use with live looping the following year. Analysing this video, along with other videos of this composition available on social networks and my personal recordings of this composition, from the first rehearsals to the live performances at festivals and concerts, it was possible to observe that: 1) experimentation with percussion instruments did not advance to the concerts; 2) the part of the arrangement that includes a modulation in the harmony was never performed in any concert; and 3) audiovisual recordings of this composition had minimal impact compared with its live performance. The absence of percussion in live performances may be due to the fact that there were already many problems with feedback to overcome, leaving only the option to reproduce percussive sounds with the voice or body of the guitars. It is also a matter of fact that the technical acuity in the world of percussion instruments is not firmly established for me as a composer and guitarist. The fact that modulation was left out of the arrangement of the song may be related to a lack of contentment and satisfaction with this idea, but this is still an unknown.

With regard to the lack of impact of the audiovisual recording of this composition, it may be due to the quality of the recording, from the visual production to the low quality of sound and image, but, based on the live experience, I believe it to be directly linked to the fact that that in live-looping compositions where the musical idea is shorter, more "repetitive", there is better interaction with the audience in a live performance compared with the audiovisual, so the audience reaction allows for greater "public-artist" engagement.

Regarding the idea of "repetition" and "shorter musical idea" when it comes to live looping, the examples of the compositions *Daqui a Você* and *Inês e a Ilha* (available on YouTube and other social networks) demonstrate the situation with the audiovisual well. In *Daqui a Você* ("From Here to You"), which has the structure "Introduction (C') ABABABC'AB Final (C)", the arrangement's introduction alludes to the melody that occurs at the very end ("C"), which is also revisited within the song's structure before returning to conclude the song. This introduction is brief, but it draws attention to an idea that is not continued but instead goes from there to build the loop that will be used in the "A" part of the song, repeating itself a few times in the arrangement. But what attracts the audience's attention – as mentioned by several people, both musicians and non-musicians – is that the soundscape present in part "A" is completely different from the other parts of the song and, having in loop the sound of the guitar, bass, and drums, causes a clear sense of an arrangement performed by more people, which leads to an interesting audiovisual effect.

This composition was performed as an instrumental in the video and later featured on the Modulatus album as a song, as lyrics were added in 2018, having been shown at festivals in Portugal, Brazil, and Spain, always generating a great response from the public, whose feedback was duly recorded. With the song *Inês e a Ilha* ("Inês and the Island"), the use of live looping allowed adding one more part to the composition, which was added to the end of it, but no less important, as the overlapping of melodies works as an "answer" to the previously presented structure.

This composition has more of a jazz combo character and was until then an instrumental piece for acoustic guitar. that surprises because the loop appears at the moment when the music seems to have ended, which was very noticeable in several concerts, as the audience usually applauded before the beginning of the Loop, thinking the song had reached the end. This song features on the instrumental album that I'm currently recording (at time of writing) and it is very interesting to see how the melody overlay, now being shared between the guitar, saxophone and piano, connects with the composition in general, and sounds very good.

Final thoughts and remarks

Analysing the creative process, the use of live looping in the performance, and the way in which all this interacts and reacts with the composer and the audience, was a great opportunity to observe the artistic concept "externally" and understand a little more about what it intended to say and what it had managed to say so far, but mainly to assist in the development of future projects. Being able to see and hear oneself with critical but mainly analytical eyes brought some wisdom that is a precious resource with which to conduct the recording of my album.

The review of my equipment showed that the number of setups with the live looping feature has been inversely proportional to the mastery of these technologies, because as I acquired equipment, the time in which to study and explore them as much as possible became limited, which, despite not having harmed performance, is still a factor to be re-evaluated. It was also possible to see that an optimal strategy had not been found for establishing an intersection between the guitar and electric guitar setups, in terms of versatility – using both in the performance – and also practicality, to make it possible to transport the equipment without major difficulties. The voice effects processor that is part of the most used setup currently puts performance at risk, because of the feedback, the fact that it does not offer loop recording on separate tracks, and also the amount of footswitch required by this equipment that can force the musician to use incorrect trigger combinations in live performances, especially during improvisation situations.

Analysing the pedal chain and revisiting the notes and recordings of rehearsals, etc., served as a reminder of how complicated it is to deal with signal flow and that the more connections there are, the more complicated it will be to deal with noise. This has served as a "what not to do again" lesson, until such time that I have lots of financial resources with which to hire audio engineers and buy all the highest quality equipment possible. That said, a good proportion of the noise problems was clearly resolved with the reduction in equipment, which did not harm the performance or detract from the artistic concept. However, I recently acquired a condenser microphone and some instruments typical of indigenous and Afro-Brazilian culture, which are in the experimental phase of being integrated into the setup, in order to connect with the initial idea of the project, which until now has only been in rehearsals.

In the context of the pandemic (ongoing at time of writing), with the demand for audiovisual productions and concerts in livestreaming format, the quality of the signal flow has become an even more important issue, as I had found that all my equipment is liable to generate noise, some more than others, and that even minimal noise will be considerably enhanced by audiovisual recording and reproduction equipment, which also create latency problems that make live looping performance very difficult. This is one of the investigations I am currently undertaking in order to adapt my concept as soon as possible to the new ways of presenting art to the world.

Regarding the use of live looping in a group performance situation, until now it has been a big unknown, as, even with great feedback quality, it relies on the musical perception of different people acting together and interacting with the pulse of the music in their way, considering that the loop is minimally "time-shifted" with each execution. In my experiments thus far, there have been a few instances that went well, and in the many times when it didn't work, I dealt with it by quickly stopping the execution or deleting the loop so as not to compromise performance. But this is another subject of studies to be shared in the future.

Links

Akiré (vídeo): https://youtu.be/_AERFOM0DdY
Daqui a você (video): https://youtu.be/99luFAjOfAs
Inês e a ilha (video): https://youtu.be/U6sAu1SD9ow
Concerto Iury Matias Portugal 2018 (video): https://youtu.be/16oXbcTNlXU
Concerto Iury Matias Brasil 2020 (video): https://youtu.be/WGUjQ4ReFFk
Concerto Iury Matias 2021 (vídeo): https://youtu.be/1SecDjS7r8A
Álbum "Outro Tempo" (audio): https://open.spotify.com/album/6woFfPxzzh9PCF0qeU2fqT

4
VIOLA SERTANEJA AND LIVE LOOPING IN THE PERFORMANCE OF ANTONIO MADUREIRA'S REPENTE

Erik Pronk

Introduction[1]

Live looping is characterised by the cyclic reproduction of audio clips. These clips, also called loops, are recorded in real time using any device that allows them to be stored and played repeatedly. In addition to this characteristic, there is also the possibility of overdubbing the loops, forming several layers. In this way it is possible to create ensemble-like textures from a single instrument, combining several layers of audio recorded in real time (or "live"). The live looping provides, in this way, an expansion of the possibilities of solo performance.

My interest in live looping came from my experience with the electric guitar, where it is common to use effects pedals and other types of peripherals. It is also common to see electric guitarists using loops in their performances. However, my first experience with live looping took place in the first year of my PhD at the University of Aveiro,[2] at the Live Looping Laboratory – LoopLab – coordinated by researcher Alexsander Jorge Duarte, a space for research and practice promoted by INET-md,[3] in the Department of Communication and Art (DeCA). From this participation, I was able to deepen my knowledge of the use of the technology of loop pedals.

This contact with live looping motivated me to combine the acoustic universe of the *viola sertaneja*[4] with amplified sounds, and thus explore the possibilities of arrangements using layers of recorded sounds. The *viola sertaneja* is a plucked-string instrument typically used in Brazil. It is known to have been present in Brazilian territory since the 16th century, having arrival with Portuguese colonists. The use of the viola has been associated with several musical traditions of the Brazilian people, such as the *Caipira* music of

DOI: 10.4324/9781003154082-6

central south-eastern Brazil, hence the instrument is also commonly known as the "*viola caipira*".[5] It is also present in an art of improvised singing called "*repente*", typical of north-eastern Brazil. The viola currently features in several segments of Brazilian music, and since the 1980s, it has found its place in educational institutions.

Regarding the search for repertoire to use with the loops and the *viola*, it occurred to me to explore the very familiar repertoire of Armorial music. This music belongs to an artistic movement that started in 1970 and became known as the Armorial Movement, focusing on various artistic branches with the interest of producing art with its roots in the cultural traditions of north-eastern Brazil, particularly related to the states of *Pernambuco, Paraíba* and *Rio Grande do Norte*.

The term "Armorial" was chosen for the movement's name by its leader, Ariano Suassuna (1927–2014). The word comes from a catalogue or collection of heraldic weapons or coats of arms, a practice started in Europe in the 12th century. In the context of the Armorial movement of the 1970s, as an allusion to emblems and coats of arms, the term evokes the symbols of the people, the roots of their identity, that would be present in Iberian, indigenous, and African heritage. The term Armorial came to be used by Suassuna and other participants of the movement to designate the elements identified by them as being representative symbols of Brazilian culture, especially the northeastern.

The choice of the Northeast as geographic-symbolic delimitation is due to the fact that Armorialists consider that in that region the modern influence has been introduced more slowly in the modernising process. Especially in its interior, the values, the way of life, customs and practices of the people would better express the "essence" of being Brazilian[6] (Santos, 2015).

One of the musical practices that influenced the creation of Armorial music was the "*repente*", an art form based on the singing of improvised verses. "*De repente*" means "suddenly" in Portuguese, making an allusion to the verses, improvised during the singing. This music is usually performed by a pair of singers accompanied by the *viola sertaneja*, but it can also be accompanied by the rebec or the tambourine, when it is also known as "*embolada*". The designation "*viola sertaneja*", chosen for referring to the instrument in this text, is justified by the adoption of the name by the participants of the Armorial Movement in reference to the *violas* of the *repente* singers.

The term "*sertaneja*" alludes to "*sertão*", which is one of the four subregions of the Northeast Region of Brazil and the largest in terms of territorial area. The culture and the imagery of the *sertão* area inspired several artists of Brazilian realist modern art movement. It is from where, for example, works of literature, such as O *Sertanejo*, from 1875, written by José de Alencar; *Os Sertões*, by Euclides da Cunha, published in 1902; and *Grande Sertão: Veredas*, by Guimarães Rosa, from 1956, get their title. Thus, the term "*sertaneja*" associates the viola with the culture of that region. The frequent use of the viola in the *repente* contributed to the association of the sound of the instrument with this musical genre, as heard in the music of poets like Diniz Vitorino, Manoel Xudu, and Ivanildo Vilanova.

The rhythmic formula used in the accompaniment of the *repente* is called "*baião*" or "*baião de viola*", which is performed as an instrumental interlude by the *repente* singers between the verses. The *repente*'s *baião de viola* inspired the consolidation of the Baião as a musical genre, which was disseminated in the 1940s through the work of Luiz Gonzaga and Humberto Teixeira. Luiz Gonzaga's basic instrumental ensemble, however, did not include the *viola*, but the accordion, the zabumba, and the triangle.

Due to the importance of the *repente* poets as spokespersons for the people and their use of the *viola sertaneja*, the instrument became an emblem of North-eastern culture and its use inspired Armorialists, in particular Ariano Suassuna, who invited Antônio Madureira to perform and research the instrument in the artistic activities of the movement. In addition to composing works for the group, Madureira was also responsible for the performance of the *viola* in the *Quinteto Armorial*, whose instrumental set also included the *marimbau*, the *pífano* (and the flute), the rebec (and the violin), and the guitar.

Ariano Suassuna would have known the *marimbau* through a tin *berimbau* player and discovered its name through research in the books of travellers who went to Brazil in the 19th century (Aloan 2008: 20). According to Aloan (2008), the name *marimbau* can also be understood to be an amalgamation of the names of other two instruments: the *berimbau* and the marimba. The *marimbau* used by the quintet was adapted by the artisan João Batista de Lima, who replaced the cans with a wooden body and used two strings supported on two guitar tuners. The instrument is meant to be played with a kind of drumstick over the strings with one hand while a cylindrical glass is used by the other to obtain different heights (Aloan 2008: 22).

The rebec is the medieval predecessor of the violin, the use of which was also made by singers of *repente*, for example Cego Oliveira, while the *pífano* is an adaptation of European popular transverse flutes also known as "*pífaro*" or "*pife*". The creation of the *Quinteto Armorial* was motivated by the use of these typical instruments of North-eastern culture. The set of instruments of another group related to the movement, the *Orquestra Armorial*, favoured predominantly symphonic instruments. Suassuna considered the sound of the orchestra to be somewhat "europeanised" and idealised another type of ensemble, formed by instruments such as the rabec, the *pífano*, the *viola sertaneja* and the *marimbau*.

The establishment of a compositional style and the use of popular instruments would be consolidated with the formation of the *Quinteto Armorial*, especially with the contribution of the composer Antônio José Madureira.

Despite the use of such instruments as the violin and the flute in the quintet, the difference was in the way of playing, in the way of popular artists, together with the reproduction of the timbres of their instruments, even when using equivalent "classical" instruments.

Development of the study of "Repente" with live looping

The work "Repente", named after the homonym art of improvised singing, was composed by Antonio José Madureira in 1970. The piece was recorded on *Quinteto Armorial*'s first album, released in 1974, called *Do Romance ao Galope Nordestino*. The original instrumentation includes flute, guitar, *marimbau*, *viola sertaneja*, and violin. The version presented in that recording served as a reference for the development of this study and a model for the transcription. Another version by the composer himself was recorded on the first *Orquestra Armorial* album, released in 1975, and presents some differences in structure, mainly in instrumentation, which includes a string quintet (two violin parts, viola, cello and double bass), percussion (box, cymbal, kick and triangle) and two flutes.

The main melody of *Repente* is based on the structure of *sextilha*, a poem of six verses, which corresponds to one of the forms typically used by *repente* singers. In the recording of the *Quinteto Armorial*, this melody is initially performed by the *marimbau* and then re-exposed by the *viola sertaneja*. In addition to the metric structure of *sextilha*, the melody is built on a scale called *modo nordestino*, or "north-eastern mode". This scale consists of a major scale, where the seventh degree is lowered by a semitone and the fourth degree is sometimes increased by a semitone.

A hypothesis defended by Silva (2005) is that this North-eastern mode has its origins in the process of musical education offered by the colonizers to indigenous peoples, where the liturgical modes,[7] would have been incorporated into the musical practices. Siqueira (1981, as cited in Silva, 2005) explains that the formation of this artificial north-eastern mode may have been

TABLE 4.1 Scale of D Major and the North-eastern mode of D

			Scale of D Major:			
I	II	III	IV	V	VI	VII
D	E	F#	G	A	B	C#
			North-eastern mode of D:			
I	II	III	IV	V	VI	VII
D	E	F#	G (or G#)	A	B	C

a consequence of the medieval practice of avoiding the tritone, called "*diabolus in musica*",[8] lowering the seventh degree or increasing the fourth. This scale is not, however, used exclusively in North-eastern Brazil: Tacuchian (1994–95, as cited in Silva, 2005) points to the use of this scale also in the work of Béla Bartók and in other folk melodies found in Central Europe and Asia.

The metric and melodic aspect combined with the rhythmic element of baião de viola are characteristic of several works of Armorial music such as Repente. The sections of Repente can be represented by the following sections: Intro – A – B – A' – C – A" – D – B – A'" – Coda. The introduction (Intro) of *Repente* is characterised by the entrance of the instruments one by one, forming overlapping layers with their parts. It starts with the guitar playing a figure based on pedal notes with the rhythm of the *baião de viola*. This part is followed by the *viola sertaneja*, with a cyclic accompaniment motif in arpeggios. The *marimbau* appears next with repeated notes. The last instrumental layer of the introduction consists of long notes performed by the violin and flute. It is important to emphasize that the appearance of musical material in the form of successive layers is also an analogous feature of live looping, which made it possible to adapt this specific music for the purposes of this work.

In section A, the main melody is presented by the *marimbau*, while the guitar and *viola* continue with an accompaniment similar to that of the introduction. Section B introduces another melody in canon[9] between violin and flute, accompanied by the guitar, viola and *marimbau* on a pedal note.[10] In the first re-exposition of A (A'), the main melody is presented by the *viola*, with accompaniment from the *marimbau* using a pedal note added to a guitar accompaniment of arpeggios, which also creates a melodic line in counterpoint to the main one. In section C, a new canon melody appears between guitar and viola, along with the accompaniment of the *marimbau* on a pedal note. Still in this section, the same thematic idea is repeated in canon by the violin and flute, accompanied by the *marimbau*, guitar and *viola sertaneja*. After section C there is the second re-exposure of A (A"). This time the main melody is presented by the violin, and a second melody in counterpoint is performed by the flute while the other instruments provide the accompaniment. A new section (D) presents a melody phrase in thirds between the guitar and the *viola sertaneja*, while following the accompaniment of a *marimbau* pedal note. After the re-exposure of B, section A is presented for the last time (A'"). The main melody is doubled by the violin and *marimbau*, while the flute plays the melody of section C in overlap. The accompaniments of the guitar and the *viola sertaneja* are based on the *baião de viola*. The work ends with a Coda, presenting a melody on the violin and flute from the thematic material of A. The accompaniment of the guitar, *viola sertaneja* and *marimbau* follow based on the pedal note and the *baião de viola*.

Therefore, it is possible to list the following compositional characteristics: the metric structure of the main melody based on the sextile; North-eastern scale; the rhythmic figure of the *baião de viola*; the presence of the pedal note; melodic duet constructions; cyclic musical motifs; successive formation of layers by the various instruments; and the use of imitative processes in the melodies (canon).

Methodology

The methodology adopted in this research is based on artistic practice, which is intrinsic to the research process and consists of the creative experimentation with the performance of *Repente* using the *viola sertaneja* and live looping. The research strategy involves the following tasks: listening to the tracks on *Quinteto Armorial*'s album; transcribing the piece to musical notation; recognising repetitive material (cyclic motifs, and cyclical phrases); transcribing excerpts to the viola; creating a project in multitrack audio editing software; studying the loop pedal functions; elaborating the arrangement; creating a musical notation for the pedal controls; and editing the final version in music score.

The process of listening to the audio track was done with the use of a digital version of the audio file through editing software, which made it possible to manipulate various parameters, such as adding markers onto the audio tracks, cutting, copying, and pasting sections of audio, and experimenting with different transpositions of the frequencies (pitch shift) and with different *tempi* (time stretch). The use of these resources facilitated the task of identifying notes for transcription, in addition to helping with rehearsing together in "play along" mode with the audio.

The transcription to musical notation of the original version in score-editing software was done based on listening to the audio. The original score of the work is not in a legible condition, as it was one of the works of Antônio Madureira that were damaged in a flood that occurred in the city of Recife in 1975 (Andrade 2017: 168). A digitalised score of *Repente*'s version for *Orquestra Armorial* is in a book about the orchestra, produced by the Conservatory of Pernambuco in 2015. Despite some differences between this and the version of the *Quinteto Armorial*, which I chose as a reference for this study, this score was important for the review of my transcription.

An important part of adapting *Repente* for the use of loops was the recognition of repetitive patterns in its structure. These repetitive patterns were identified in:

The cyclical patterns present in *Repente* have two characteristics: they are accompanying patterns or melodic phrases that are repeated. After identifying these repeated metric models, the following step was to transpose the excerpts to the *viola*. The translation of idiomatic gestures from other

TABLE 4.2 Repetitive patterns in Repente

Intro	Guitar accompaniment Viola accompaniment Long notes on the violin and the flute (with repetition)
A	Guitar and viola Melodic phrase on the *marimbau* (repeat with variations)
B	Imitative melodic phrase – violin and flute (with repetition) Guitar, viola and *marimbau* accompaniment pattern
A'	*Marimbau* Melodic phrase – viola with guitar accompaniment (with repetition)
C	Imitative melodic phrase – guitar and viola (with repetition) Imitative melodic phrase – violin and flute (with repetition) *Marimbau* – accompaniment pattern
A"	Melodic phrase – violin with flute counterpoint (with repetition)
D	Melodic phrase – guitar and viola in thirds (with repetition)
B	Imitative melodic phrase – violin and flute (do not repeat) Accompaniment pattern – guitar, viola and *marimbau*
A'"	Melodic phrase – violin with flute counterpoint based on the theme of C (with repetition) Accompaniment pattern – guitar, viola and *marimbau*
Coda	Melodic phrase – violin with flute counterpoint (with repetition) Accompaniment pattern – guitar, viola and *marimbau*

instruments to the viola required certain processes, such as transposing the octave to adapt the notes within the viola tessitura, and the use of a wood stick for a percussive sound production. The viola tessitura comprises the notes A2 to B4 (or A4 when tuning *Cebolão* in D).[11] Some notes in the guitar part were transposed an octave higher than the viola, as in the initial accompaniment pattern. Passages of the violin and flute were transposed to an octave below, as in the melody of section B.

For the transcription of the *marimbau* part, a drumstick was used to play the viola strings, similar to the *marimbau* technique. A cylindrical wooden stick, approximately 6 mm in diameter and 25 cm long, was used to strike the strings. The production of sound with the drumstick occurs by attacking the open strings of the viola. The strings that were not intended to make sounds were damped with the fingers of the left hand.

Before experimenting with the loop pedal, a project was created using multitrack audio editing software. By this means it was possible to record the different instrumental layers separately and create loop tracks from excerpts recorded, copied, and pasted in sequence, in an experience similar to the practice of recording in the studio, including capturing the audio and editing the tracks. Using the solutions found in audio recording and editing software, I started adapting for performance. This step included the study of the loop pedal functions in order to seek solutions for the execution of the proposed arrangement. The required loop pedal functions are described below:

- Three synchronised loop tracks are needed, allowing the *Tempo* of the loop tracks to maintain the same length on the different tracks that must be played simultaneously.
- The switching order of the Recording Pedal must be set to: Recording-Overdubbing-Playback. This is necessary in cases where a second layer is recorded in a loop track, immediately after the recording of the first one.
- At two different points in the performance, the multiple tracks in execution must be interrupted simultaneously, and this is done with the operation of a single pedal.
- The function to undo the last recording of a loop track allows you to erase the last layer recorded on a certain track. In *Repente*, it occurs in order to erase layer "b" from Loop Track 3. Only the first layer will be used later for a counterpoint with another melodic material, executed as solo.

The actions necessary for the performance of the proposed arrangement are as follows:

1) Viola plays two bars of the introduction (INTRO)
2) Track 1 Recording (1R): Loop 1, Layer 1 (L1.1)
3) Track 1 Overdubbing (1D): Loop 1, Layer 2 (L1.2)
4) Track 1 Play (1P)
5) Solo with melodic theme of A
6) Track 2 Recording (2R): Loop 2, Layer 1 (L2.1)
7) Track 2 Play[12]
8) Stop all tracks simultaneously
9) Track 1 Play (1P)
10) Track 3 Recording (3R): Loop 3, Layer 1 (L3.1)
11) Track 3 Overdubbing: Loop 3, Layer 2 (L3.2)
12) Track 3 Play (3P)
13) Track 1 Overdubbing (1D): Loop 1, Layer 3 (L1.3)
14) Track 1 Play (1P)
15) Track 3 Stop (3S)
16) Solo with thematic material of section D
17) Track 2 Play (2P)
18) Track 3 undo (3UD)[13]
19) Stop all tracks simultaneously
20) Track 3 Play (3P). Solo of thematic material of section A
21) Track 1 Play (1P)

22) Track 3 Stop (3S)
23) Solo with thematic material of the Coda
24) Track 1 Stop.

A musical notation model developed to represent the arrangement was designed to clarify the resultant sound of the various audio layers generated by Loop Station, together with the indication of pedal activation in a rhythm line. The acronyms chosen to represent the different actions of the loop pedal are: 1R; 1D; 1P; 1S: Track 1 record, dub, play, stop, respectively (the same for 2R; 2D; 2P; 2S; 3R; 3D 3P; 3S); 3UD: Track 3 undo; SS: synchronised start/stop. The representation of the sound material recorded and reproduced by the loop pedal is shown in the score in instrument lines.

Setup

The instrument's audio was captured using a microphone connected to the device's (loop pedal) input. During the first experiments with the loop pedal, headphones were used for audio monitoring, connected to the device's output. The first experiences with the inclusion of amplification columns created the problem of the sounds transmitted by the speakers being picked up by the microphone and undesirably incorporated into a loop track.

Regarding amplification, the preference was to favour the acoustic sound of the instrument, amplifying only the audio signal from Loop Station, thus creating a dialogue between the acoustic sound of the instrument and the amplified sounds of the loop pedal. The audio capture, which was initially done with a microphone in a fixed position on a pedestal, was changed for a lapel-type microphone, as the body movements needed to activate the pedals generated an unbalanced sound capture, due to the variation in the distance from the sound source (instrument) and pickup (microphone in the fixed position).

Regarding body posture, a strap, fixed at two points, was used to support the instrument on the body, which made it possible to perform standing up, maintaining the stability of the instrument in relation to the body movements necessary to activate the pedals. The setup used can be illustrated by the following diagram:

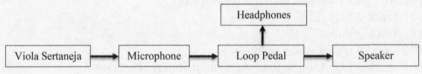

FIGURE 4.1 Setup used in the performance of Repente with live looping

Analysis and discussion

The musical excerpts transformed into loops were chosen in order to maintain the original structure of the work, with no significant alterations or fragmentations of the musical material. In the version proposed for performance with loops, the presentation of the various layered parts takes place in a slightly different way to the original, as the layers appear successively, while in the original version they are presented simultaneously. Examples are found in section B and section C. In this way, the proposed version for live looping brings a progressive increase in instrumental texture, since with each new exposure of a certain section, a new layer is added. This can bring a musically interesting result, as the new version is made based on the sound of a single instrument, without the timbre variety of the *Quinteto Armorial* ensemble.

The three loop tracks used in the arrangement appear either separately or combined. Therefore, it is necessary to use an equipment that allows the creation of three distinct loop tracks, which can be operated independently. In *Repente* we have an example in section C, where Track 3 is repeated every six exact cycles of Track 1. For the beginning of each cycle of Track 3 to start exactly at the beginning of six Track 1 cycles, it is necessary that the tracks are accurately synchronised.

Types of loops used in "Repente"

From the observation of the musical material contained in the loop tracks created in the arrangements, it was possible to identify two distinct types: Loops of Cyclic Motif and Loops of Whole Phrase. Loops with cyclic motifs are shorter – they consist of figures that serve to compose an accompaniment – and are present in Track 1, made up of three layers in total. Whole-Phrase Loops are more extensive – they have a sense of syntactic musical unity, with a broader range of elements than Cyclic Motif Loops – and are present in the phrases of themes B and C. The REC Pedal Activation (RECORD) can be synchronous or asynchronous to the execution of the notes of the instrument. In Synchronous activation, the REC Pedal is activated simultaneously with the execution of the first note of the phrase or cyclic motif. The playback of the loop tracks occurs immediately, that means that the recorded loop track is played right after its recording.

Conclusions

Through this study case it was possible to observe the possibility of using loops of real-time recorded sounds in the arrangement of a musical work of Armorial music, based on the piece *Repente*, composed by Antônio José Madureira for the *Quinteto Armorial*. The work was chosen based on the need to find appropriate repertoire for live looping. The use of loops made possible to recreate the texture of the original version through the performance of a solo instrument, the *viola sertaneja*.

The resulting arrangement, however, presents some differences from the original: the different layers, which appear simultaneously in the original, appear successively in the version proposed by this study. However, this result was artistically interesting in the sense of adding more variety to the new version, since it uses the timbre of a single instrument, while in the original version the work features a five-part ensemble.

The arrangement was characterised by the combination of acoustic guitar sounds with amplified sounds stored in the loop pedal. The choice of the viola is related to the genre of *repente*, which inspired the composition of Madureira's work, as well as being part of its original instrumentation. The methodology adopted was based on the artistic practice of the author of this study and the process of recreating the work, which was made possible by the presence of repetitive patterns in its compositional structure.

The adaptation of the work through the use of loops was initially made with the aid of audio editing software, with the audio being recorded and the loops created manually with the cutting and editing tools. To create the arrangement for a live performance, it was necessary to experiment with the programming possibilities using the loop pedal. The performance preparation process included, therefore, the study and practice of the loop pedal tool, in order to achieve success in synchronising the sounds played by the instrument together with the recorded sounds.

The research thus presents a new performative possibility of this repertoire through the interconnection of the *viola sertaneja* with technological interfaces.

Notes

1 Throughout the text, the term "viola" refers to a guitar-like Brazilian instrument.
2 Pronk 2021.
3 INET-md: Instituto de Etnomusicologia, Centro de Estudos em Música e Dança.
4 The *viola sertaneja* is an instrument widely used in my region of origin, in the northeast of Brazil.
5 It is worth noting the fact that several names could be used as a reference to this instrument in Brazil; however, the relationship between instrument names and characteristics is sometimes ambiguous, as different names could refer to the same type of instrument, while the same name can refer to instruments with different characteristics.
6 However, it is important to note that this distance from foreign influence did not occur only in the Northeast region of Brazil, as defended by the Armorialists. It also took place in other regions of Brazil, with an emphasis on the Midwest and the North, but also in some areas of the Southeast, such as the north of Minas Gerais, the Vale da Ribeira in São Paulo, in addition to other regions of Espírito Santo.
7 The liturgical or ecclesiastical modes were used in music especially in religious celebrations as plainsong. They received Greek names, alluding to the Greek regions

where these modes were supposed originated, for example, Doric, Phrygian, Lydian, Ionian, Aeolian. Mixolydium is a mixed mode between Lydian and Doric modes. The use of the Ionian and Aeolian modes was established in the late Middle Ages as the major and minor scale, respectively.
8 The interval formed by the overlapping fourths of the scale, for example: C-F; D-G; E-A, form a perfect fourth, with two and a half tones. The F-B interval consists of three tones (or tritone). Medieval musical rules did not allow this dissonance, which became known as "*diabolus in musica*". To correct the tritone, one could increase the F a half tone up to F# or lower the B half tone down to the B flat.
9 The canon is a polyphonic technique in which one or more voices imitate the melodic line presented by a first voice, entering each voice, one after the other, repeating exactly the same melodic line as the first one.
10 The pedal (or drone) is a prolonged or repeated note on which different melodic lines or harmonies are built.
11 Considering the Portuguese numbering of octaves (Med 1996: 266), where the central C of the piano is C3.
12 Once the order of operation of the recording pedal is set to the order Recording-Overdubbing-Playback, it is necessary to press this pedal a second time to stop the recording of an overdub track.
13 In some equipment it is possible to assign this command to a pedal, for example the play pedal of Track 2.

References

Aloan, Rafael Borges. "A organologia e a adaptação timbrística na música armorial" (Final Paper, Universidade do Rio de Janeiro, 2008). https://econtents.bc.unicamp.br/eventos/index.php/eha/article/view/3776.

Andrade, Francisco. "Quinteto Armorial: timbre, heráldica e música" (Master's thesis, Universidade de São Paulo, 2017). https://www.teses.usp.br/teses/disponiveis/31/31131/tde-13122017-112348/pt-br.php.

Pronk, Erik de Lucena. "Do romance ao loop nordestino: viola sertaneja e live looping na música armorial" (Phd Thesis, University of Aveiro, 2021). https://ria.ua.pt/handle/10773/33121.

Santos, Nivea Lins. "O Galope Nordestino Diante do Parque Industrial: o projeto estético do Quinteto Armorial no Brasil moderno" (Master's Thesis, Universidade Estadual Paulista, 2015). https://repositorio.unesp.br/handle/11449/134153.

Silva, Vladmir. "Os Modos na Música Nordestina". Piano Class – Revista de Música e Artes (2005): 1–16. https://pianoclass.com/pt-br/os-modos-na-musica-nordestina/.

Discography

Repente. Composer: Antônio José Madureira. In: Quinteto Armorial – Do Romance ao Galope Nordestino, 1974. LP.

5
QUASITUDE
The processes and methods of the composition work for xylophone and live looping

Helvio Mendes and Samuel Peruzzolo Vieira

Introduction

Quasitude (2020)[1] is a work composed in the scope of a PhD research project that intends to explore the sound possibilities of the xylophone from the perspective of potentiating it, both in the production of new works and in the artistic creations for the solo instrument. The work was developed through the sonic exploration of the instrument using technologies associated with live looping.

The performer was responsible for all the research on instrumental exploration and sound processing – through electronic platforms – of the xylophone. The composer, in turn, modelled the samples, conceptualised the narrative and the character, and set the sound material in a musical work.

The collaborative process between composer and interpreter sought to be comprehensive and inclusive since it embraced complementary perspectives, namely improvisation, free experimentation, technical research, aesthetic valuation, etc. This phenomenon already points to a constructivist perspective that seeks "to distinguish without disjunction, to associate without identifying or reducing" (Morin 2005: 15). Therefore, the conflict between composition and interpretation shifts out of the hierarchical relationship, towards its dilution. Such displacement is evidenced by the fact that the composer's supremacy has been, over the years, gradually deconstructed, with much encouragement from the performers themselves (Domenici 2010: 1144). In this sense, the need to include forms of collaboration that seek to expand and potentiate collaborative practices becomes evident. *Quasitude* seeks to contribute to this effect, serving as an experience report in which both composer and performer add complementary knowledge about creation, interpretation, performance, and technologies.

DOI: 10.4324/9781003154082-7

Live looping (LL) is a technology used in musical performances that uses loops recorded in real time. With the advent of the Loop Station pedal and hardware/software apparatus that can be installed on laptops or mobile devices, a wider spread of the use of technology is observed (Duarte 2015: 11). LL is typically applied in the enrichment of instrumental music, especially in solo situations. By introducing repeated rhythmic patterns or creating melodic motifs, the LL, combined with percussion instruments performed in real time, seeks to create musical narratives and expand possibilities (Mendes et al. 2018: 6).

Based on the use of LL as a compositional tool, this work discloses the laboratory reports of experimentation and creation that resulted in the work *Quasitude* for xylophone and live looping, as well as the collaboration between composer and performer.

The methodology adopted generally went through the following steps:

1) the research process on the exploration of acoustic sounds of the xylophone and later combinations with digital filters from the use of DAW's (Digital Audio Workstation) made in the Live Looping Lab (LoopLab/ INET-m);
2) a critical approach to current practices in the relationship between composer and interpreter;
3) processes of conception, composition and notation of the work developed by the composer from the sound material offered by the interpreter.

In this sense, we intend to shed light on the processes of composition and interpretation of *Quasitude*, with the intention of structuring new possibilities of sound exploration in the xylophone, with a view to encourage future similar artistic creations.

Xylophone

In the occident, the xylophone as a standardized instrument (with bars displaced as a piano and with resonance tubes) spread in different musical contexts and genres since the beginning of the 20th century. Cahn (1979) considers that between 1880 and 1925, there was a "golden age" of the instrument, linked to the profusion of xylophonists who performed as soloists in different orchestral formations and chamber groups. In the second half of the 20th century, with the growing number of performers dedicated to the marimba and others attracted by the mechanical improvement of sustaining notes on the vibraphone, the xylophone lost its prominence as a solo instrument (Strain 1995: XI). This resulted in a reduction in the number of new works for the instrument within contemporary classical music compared with those for marimba and vibraphone.

Since the early 1970s, the xylophone has been researched from historical and application perspectives of the instrument in Western music (England 1971; Eyles 1990), with studies on documentation and recordings between 1877 and 1929 (Cahn 1979), musical life and musicians' participation in its dissemination (Lewis 2010; Goto 2013), and improvisation in jazz and popular music in the 1920s and 1930s (Singer 2017). However, there is a gap in approaches to solo repertoire and exploring its sonic possibilities.

Yet during the 1970s, there was increased musical production in the context of ragtime. One hypothesis for this increase relies on the work of percussionists John Beck and William Cahn,[2] which revived, among others works, the G. H. Green's solo xylophone[3] works, creating new arrangements with marimba accompaniment (Singer 2017: 1). However, although of paramount importance, this movement did not give rise to further reflections on the study and practice of the instrument outside this context seeking to understand what perceptions and possibilities might emerge from this field.

Live looping

Given the context presented of the instrument, one of the possibilities for enhancing sound sources on the instrument was the use of live looping. For this, in the scope of academic research, a search for works written for xylophone and LL was carried out and the minimal results were found.

In the expectation of filling this gap, contacts were made with contemporary classical composers, in which the avant-garde of experimentalism was predominant in its compositional ideals, with the interest of realising collaborative works, from the idealisation of the work to its unfolding in the compositional and performative process.

The performer's interest in understanding the use of LL and its possibilities for xylophone sound expansion compelled him to involvement in LoopLab/ INET-md (Live Looping Laboratory of the University of Aveiro). The research went through the process from the perspective of a) understanding how the existing LL platforms work; b) obtaining basic knowledge of audio equipment, microphones, and sound flow; c) performing basic practices on LL equipment; and d) experimenting and creating possibilities of performative practices with xylophone and LL.

The result of this immersion period culminated in the composition of the work *XyLoops* (Mendes, Traldi, & Duarte 2018), in which Traldi conceived the structure of the work and Duarte and Mendes cooperated in the technological and instrumental aspects, respectively. The composition experience and the resultant sound material led to the composition of *Rastros #2* (Traldi & Barreiro 2018) for xylophone, symphonic band, and electroacoustic sounds (in real time and LL). Similarly, the performer took from this experience the application of electronic processing on the different timbres of the xylophone, mainly when applied to the free solos in the work of Traldi and Barreiro (2018).

During the investigation and performative practices, the LL device expands the xylophone beyond its typical setup, in which the interaction is between the instrument, the sheet music and the mallets. This approach requires not only more gestural actions from the performer but also requires an increase in skills to create different overdubs during the performance. In this sense, the LL performance differs from other electronic works. The most often is fixed media in which the performer doesn't need to trigger the device, using the electronic accompaniment as a backing track provided by a tape, CDs, or other devices.

The LL also differs from other more complex electronics, such as live electronics with Max/MSP and Pure data software, that can recognize gestures or sound blocks to trigger elements. The here used LL relies on the performer for the real-time creation of different overdubs.

The logistics of using electronic platforms were another aspect considered by the researcher-interpreter. According to the demands of the concert, the set up can be assembled at different levels of complexity. Therefore, the interpreter must have a mastery of the LL electronic platforms so that they can be adjusted according to the demand. In other words, a performance in a studio or on stage can be assigned with different setup and configuration logistics for the same work. It is up to the interpreter to know how to choose the tools that provide better cost-effectiveness without harming the execution of the work.

Composer-performer collaboration

The composition of a musical work based on collaboration presupposes different stages and perceptions. Among the many common examples are: the stages for the elaboration of the work as a product of an artistic creation, the synergistic relationship that the composer-performer collaboration imposes, and the various phases and manipulations that the material undergoes after its initial concept. However, not all collaborations are established from a neutral perspective as far as background knowledge is concerned. They can be situated either in a "foreign" context, in which the composer does not play the instrument for which they write, or in a "domestic" one, in which the composer knows the instrument they are writing for, as well as its singularities, techniques and repertoire. The latter tends to establish a very close relationship between expressivity and instrumentality since

> The composer-performer's creative procedures contrast to the other type of composer on the process of communication and self-evaluation, allowing him/her to better understand the other performer's point of view and

the advantages of writing for his/her own instrument. The knowledge of idiomatic writing allows the composer-performer to reflect and choose the compositional intention based on empiricism [and] the very nature of his/her own language to speak directly to the performer.

(Peruzzolo-Vieira 2015: 650)

For this research, both the performer and the composer are classically trained percussionists. Because of this, the work went through a creative process in which the composer composes from prior knowledge. In order to come up with a work that did not lose the essence of the sound exploration of the xylophone and the performer's interests, the compositional process included aspects from both sides. In addition, the use of unconventional materials and gestural approaches in the form of extended techniques gives the work a capacity to free itself from the technical-performative repertoire in favour of a language that seeks to move away from technical-expressive patterns proper to the traditional repertoire of the instrument.

The exchange of influence that composer and performer exert on each other generates new working dynamics, where ideas are consolidated in the form of compositional stages. Due to the peculiar aspects involved – instrumental techniques specific to each performer, extended instrumental techniques, working flow, project nuances, etc. – the relationship implies a close and regular contact. This contact, in turn, ends up opening to the composer a field of possibilities often unique and exclusive to a particular performer.

The working area, enriched by the performer's information, allows the composer to formulate and deduce compositional instances that, without direct contact with the performer, would be quite difficult to reach. It is worth mentioning that such a relationship does not necessarily imply a subversion of roles. Since the creation departs from a cooperative state, the working dynamic resizes the process of creation and construction of the performance, thus, in some way, a method of assisted composition of the musical work occurs.

Laboratory (three phases)

The study carried out in the laboratory went through three phases, designated as Phase 1 – Experimentation, collection of generating sound materials performed by the interpreter; Phase 2 – Implementation, the processing of these materials within the work elaborated by the composer; and Phase 3 – Combination/intersection of acoustic and electronic media investigated by the interpreter.

Experimentation – data collection that generates sound materials

The sound exploration on the xylophone began with experimentation using unusual materials directly on the instrument. This research phase follows the concept of *instrumentality*, in which Hardjowirogo (2017) states that the use of an object to produce sound does not depend on whether it is designed to be or to be used as a musical instrument but on the context of its use.

The experimental materials resulted in a diversity of sonorities unusual to the instrument's usual practice, which was subsequently catalogued in a database.

For this project, among several unusual materials tested, those that require the movement of friction to produce sound were chosen to emulate a reco-reco (idiophonic scraper). The following materials were selected: straw broom, marble, violin bow, a corrugated tube of different diameters, and ping-pong balls.

Carlos Stasi, a famous instrument researcher, defines reco-reco as the generic name given in Brazil to percussion musical instruments whose main characteristic is to have a surface with high and low reliefs (usually in the shape of teeth) scraped with different objects. The author adds that such instruments, as well as the objects that scrape them, are made of different materials and take on the most diverse forms, besides being played in different positions and presenting numerous playing techniques (Stasi 2011: 19)

Based on Stasi's premise, the material experimentation and exploration focused on the acoustic sound results based on the xylophone's frame and resonators without sound processing.

Below is a brief description of the unusual materials used in this project:

Bow for a string instrument: the manipulation of bows on percussion instruments is already a recurring practice and is present in several works of the repertoire, especially for vibraphone. It presents some limitations when used on the xylophone since the capacity of wood vibration is smaller than the capacity of the vibraphone metal vibration. However, it is possible to achieve an extraordinary exploration of harmonics in the bass region of the instrument because the wood is thinner and greater length.

Straw broom: the straw broom used on the xylophone presents specific sound characteristics. The aim is to apply it with friction on the xylophone's wood in different movements: bass to high, high to bass; natural notes to accidentals; accidentals to natural notes. The glissando movement on the xylophone causes a sonority analogous to what is known in acoustics as white noise.

Corrugated tube: this material resembles the reco-reco described by Stasi (2011). It was explored by using the ends of the wooden bars to serve as the material that scrapes the tube; with a long flexible tube making it possible to scrape more than two bars simultaneously. The larger the diameter of the tube, the more the sound result will favour the low frequencies of the wood.

Marbles: marbles are glass balls that, depending on the way they are touched on the xylophone, may cause different sounds. In friction with the xylophone, the sound result can vary according to the number of balls and the friction energy.

Table tennis ball: the contact of the table tennis ball with the xylophone causes interesting sound peculiarities, and the control of its rebound makes it possible to perform the effect known in percussion as *buzz*.

Implementation – processing the materials at work

The piece contains three main sections, each featuring different compositional techniques. The first section features *layering*, in which distinct sound materials are progressively superimposed as the derivations of these materials into new ones are triggered. In section two, the work moves towards more linear writing, maintaining a clear identification with the compositional material presented previously. In this section, the approach distinguishes itself by demonstrating instantaneous sound processing. Finally, the work moves towards a closure in which new performative elements appear within a creative approach already used. In comparison, the second section (page 2) presents a more cantabile discourse than the first, more textural one. In the former, the narrative highlights an innocuous but clearly expressive language. Gradually, the solo leads to the work's climax: a frenetic gesture in the treble register repeated to exhaustion. In the coda, the narrative moves back to a contemplative state, including elements from the first and second sections.

Quasitude is written in proportional notation. Proportional or spatial notation is a visual organisation of the music score in which events are spatially arranged in the score and follow each other topographically in relation to the distance/proximity of reference points. Besides the proportional relation, the score also includes a temporal relation measured in seconds, as can be seen in the score (Figure 5.1). The measure formulas serve as arbitrary references of this proportionality, indicating metrics varying between long, medium, short, and very short performative gestures.

The choice of materials used in the composition prioritised the cohesion of the compositional narrative. Thus, the concept of compositional modelling (Pitombeira 2011) was musically represented, especially by irregular parameters such as noise, randomness, friction, and periodicity. According to Pitombeira (2011: 39), compositional modelling consists of three spheres: temporal, spatial and narrative. The first comprises the concepts of duration and elasticity. The second includes medium, material, texture, and density, and the third is context, character, situation, and gesture. The composer sought to find a creative space for each element in dialogue with the premises presented by the interpreter, concomitantly relating the modelling concept to the creation of the work.

Another fundamental element for the creation of *Quasitude* came from reading the book "Instrumento do Diabo" by Carlos Stasi (2011), in which the author coined the term *quasitude*, relating it to the verb to scrape, based on the following proposition: "the idea of the quasi-totality of an action, that is, the fact that an action implies the act of scraping is never really completed, never materialised, it is always something that, as we say every day, 'barely scrapes', almost happens" (Stasi 2011: 15). From this term, *Quasitude*'s original intent was to move away from the characteristic elements of the technical-stylistic repertoire of the xylophone. Hence, the idea of an unfinished intention underlies the piece, for it suggests an implicit intention in its way of approaching the sound. Such mindset led the composer to a phenomenological investigation about the gesture in music performance as an act of revolution and mould breaking. Therefore, the composition seeks to reveal a balance between the performer's action and the electronics' reaction. The use of the technology results in a technical-stylistic extrapolation of both the performability and the language adopted.

Below we discuss the materials used and how they were incorporated and organised within the work.

Acoustic and electronic media combination/ intersection

From the contact with the music score, the performer started to realize experimentation processes with the acoustic sound results proposed by the composer in combination with electronic sound effects. Between the acoustic sounds offered to the composer were used at the composition, corrugated tubes, table tennis balls, chopsticks, and the drumstick handle.

Initially, the score was analysed in order to identify possible incongruities related to performance issues on the xylophone, as well as the structural analysis of the work. During this process, extended technique practices essential for using the selected materials and the consideration of some possibilities of timbres with the electronic processors were already possible.

The Loop Station pedal was used from the first moment, however, the hardware proved to be limited in two crucial points for the performance of the piece. Firstly, the limitation of creating overdubs, as the work requires six overdubs with different tempos and the pedal, although having three available tracks, was not precisely capable of satisfying such a need. The second corresponds to the limitations of the effects of xylophone sound processing, as some parameters do not respond to the sounds generated by the xylophone, demanding the use of other electronic sound processors.

Experiments started following two work fronts. The first was related to solving the creation of six overdubs of different time durations, and the second was the experimentation with other electronic sound processors to

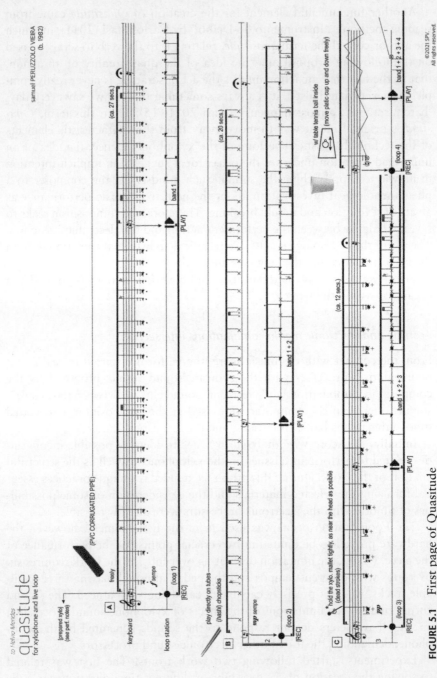

FIGURE 5.1 First page of Quasitude

Created by the authors.

choose effects on the xylophone. Ableton Live (AL) software and the GarageBand (GB) app have experimented with this process.

AL is a digital audio workstation (DAW) whose main differences compared with other similar software are linked to its possibilities for recording, arranging and sound transformation in real time. Besides having the typical functions of other DAWs linked to compositional practice in delayed time, AL also presents a wide range of tools designed for live musical composition and performance. GB is also a digital audio workstation or DAW that can be used both on a computer via the software or with a smartphone device.

The experimental work with the electronic tools achieved two different results: 1) performance with p AL without sound processing, and 2) performance with the Loop Station and sound processing in GB.

AL has a plugin called Looper, which allows the configuration and creation of overdubs in different ways, allowing the interpreter to develop various strategies for using LL. This way, with the help of a midi pedal, AL

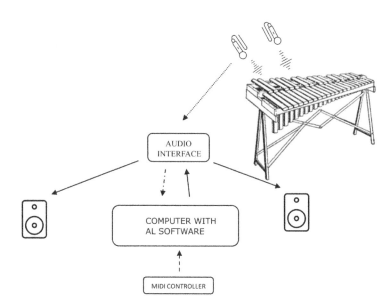

············▶ Microphone for the audio interface;
‐ ‐ ‐ ‐▶ MIDI Controller for the computer with AL software;
‐ ·· ·▶ Audio interface for computer with software;
——▶ Sound Processed by AL and send to the audio interface and then to the speakers.

FIGURE 5.2 Example of the sound flow

Image created by the authors.

managed to supply the demand of six tracks within the durations required by the composer.

GB obtained better results for the sound processing of the xylophone and, despite the simplicity of the layout on the smartphone, the results obtained with the Echo Tape, Echo Waves and Echo Warp effects were provocative and expressive when combined with the sound material generated by the xylophone. As previously shown, it was impossible to configure the Loop Station pedal to record the overdubs at the tempo suggested in the score.

Among the platforms used, AL succeeded in satisfying the needs of the work. Furthermore, when combined with the sounds generated by the xylophone, the platform's wide range of effects achieved the same results as GB.

Another factor that reinforces the use of AL for the work is the simplified assembly of the sound flow system. The sound capture of the xylophone was done through a pair of overhead directional microphones and transmitted to an audio card. Next, the audio was transported to the computer, where the overdubs controlled by the midi pedal were created. Finally, the sounds processed by the computer were conducted to the audio card and then projected to the speakers.

Conclusion

In this chapter, the composition process of *Quasitude* for solo xylophone and live looping was discussed. The results achieved in this process revealed a) new sound possibilities for the xylophone through the combination of acoustic sounds with electronic filters; b) an alternative synergy between composer and interpreter for the composition of the work; and c) compositional and interpretive potentialities in the exploration of LL.

Furthermore, the use of the LL allowed to obtain tools for composition work. The research was open to experimenting with LL on different platforms to select the one best suited to the compositional proposal. From the perspective of the relationship between performer and composer, it showed that the breaks in hierarchies can contribute to more enriching compositional constructions.

The investigation also highlighted Stasi's (2011) thoughts on the concept of *quasitude*, in which scraping is associated with an incomplete action that connotes imperfection. However, scraping on the xylophone was an enriching exploratory path in the search for the instrument's sonic expansion and generated compositional possibilities when used with the LL tools.

Finally, the xylophone sound exploration was associated with collaborative musical creation, resulting in a fusion of perspectives and intertwined experimentation and improvisation.

Notes

1 See it here: https://drive.google.com/file/d/1JH-IsELotxbqzRpJeCcN-1I6Acdivg5j/view?usp=sharing.
2 Cahn compiled a discography of each recording with the instrument and presented the book *The Xylophone in Acoustic Recordings* (1877–1929) and Becker collected scores and lesson books from various sources, including musicians who knew, studied or played with Green (Williams 2013: 118).
3 Green played classic xylophone excerpts in vaudeville theatre, as well as novelty rags and ballroom music with his own trio or with piano accompaniment and with numerous dance orchestras (Singer 2017: 3).

References

Bittencourt, Pedro Sousa. "Interpretação participativa na música mista contemporânea", *Revista Interfaces* 1, no. 18 (2013): 104–115.
Cahn, William. "The Xylophone in Acoustic Recordings (1877 to 1929)", The Percussionist 16, no. 3 (1979): 133–152.
Domenici, Catarina. "O intérprete em colaboração com o compositor: uma pesquisa autoetnográfica". Anais do XX Congresso da ANPPOM. UDESC, Florianópolis, 2010, 1142–1147.
Duarte, Alexsander Jorge. "A arte do looping. A loop station como instrumento de prática performativa musical", *Post-ip: Revista do Fórum Internacional de Estudos em Música e Dança*. 3 (2015): 9–20.
Mendes, Helvio Monteiro, Duarte, Alexsander Jorge, & Traldi, Cesar Adriano. "XyLoops – Composição e performance de uma obra para xilofone e eletrônica em tempo real (live looping)". *Revista Vórtex* 6, no. 2 (2018): 1–20.
Moersch, William. "Marimba revolution". In *The Cambridge Companion to Percussion: Cambridge Companions to Music.*, edited by R. Hartenberger (Cambridge, Cambridge University Press, 2016), 43–54.
Morin, Edgar. *Introdução ao pensamento complexo* (trans. Elaine Lisboa) 3ª ed. (Porto Alegre: Editora Sulina, 2005).
Singer, Jonathan. "Noodling Changes: The Development of Xylophone Improvisation in New York City (1916–1942)" (PhD Thesis. Graduate Faculty in Music, City University of New York, New York, 2017).
Stasi, Carlos. *O instrumento do "Diabo": Música, imaginação e marginalidade* (São Paulo: Editora Unesp, 2011).
Strain, James. "The xylophone (ca. 1878–1930): Its Published Literature, Development as a Concert Instrument, and Use in Musical Organizations" (PhD Thesis. Eastman School of Music, University of Rochester. New York, 1995).
Peruzzolo-Vieira, Samuel & Carvalho, Sara. 2015. "The composer-performer and the problem of notation in indeterminate music". Ninth Triennial Conference of the European Society for the Cognitive Sciences of Music, Manchester/UK. Escom, 2015, 646–651.
Pitombeira, Liduino. "Paradigmas para o ensino da composição musical nos séculos XX e XXI". *Opus* 17, no. 1 (2011): 39–50.
Vieira, Márlou Peruzzolo. "The collaborative process from the performer's perspective: a case study of non-guitarist composers" (PhD Thesis. Universidade de Aveiro, 2017).

6

DENSUS BRIDGE

For trumpet and live electronics
(live looping and effects)

Elielson da Silva Gomes and Alexsander Duarte

Introduction

The 20th century brought many transformations in various scientific and artistic fields (Hobsbawm 1995; Gatti 2005). In music, specifically, several aesthetic trends and artists emerged with a series of new possibilities for creation, which generally distanced them from the previous musical styles (Kerr 2012; Gatti 2005). In this historical panorama, two major changes took place in music: (1) avant-garde artists gave musical art an innovative language, breaking traditional models and bringing a new perspective to creation and listening, and (2) The dodecaphonism created by Schoenberg (1874–1951) in 1924, broke with the tonal system that had been in force since the 17th century, on which musical compositions were based. (Kerr 2012: 57). However, the greatest development of this sound revolution took place after World War II, when electronic music was suddenly developed in studios using magnetic tape, computers, and synthesisers, allowing the performance of live electronics (Holmes 2013: 34, 35; Dunn 1992).

In the 1950s composers John Cage, Edgard Varèse, and Karlheinz Stockhausen Stockhausen, based on the new possibilities of changing sound, were the first to use tape in their compositions, they changed the sounds by modifying them, and reverse combining at different speeds (Griffiths 1987: 145). From that period onwards, live looping emerged in musical performance with the musical practices of Terry Riley. Riley's experimentation with tape consisted of recording sounds by doing random operations and patching them up by changing the speed during the live performance (Meyer et al. 2014: 353). This was the technological tool used to create a live loop, although it

DOI: 10.4324/9781003154082-8

was only at the beginning of the 21st century that the BOSS company presented the first pedal loop machines. These machines operate with the feet allowing you to create and record audio and reproduce this audio cyclically (Duarte 2016: 11).

The use of electronic devices in the trumpet repertoire is relatively new – second half of the 20th century – and this electronic innovation has allowed composers to expand the trumpet's musical potential (Barth 2011). The application of these devices is intended to create specific sounds and thus alter conventional trumpet timbres (Cherry 2009). The literature on trumpet practice with the use of electronic tools, such as the loop pedal and guitar pedals (easily found on the market), has generated significant knowledge. However, the most determining factor for real learning about this new aesthetic discourse was the process of experimentation, creation and research carried out at the Live Looping Laboratory (LoopLab) at the Department of Communication and Art (DeCa) at the University of Aveiro. The LoopLab was fundamental for a deeper understanding of the process of creation and performance of the work *Densus Bridge*.

The work started from three motivations, which coincide with those that Maxwell (1996, p. 220) lists: 1) personal, 2) practice, and 3) research. Personal motivation concerns the interest in artistic improvement as trumpeters. As for the practical motivation, this has to do with the interest in using this tool in addition to its casual use, along with another technological tool, namely the loop pedal, and exploring its various possibilities. Finally, the motivation for the research concerns the need to study the possibilities that live looping (LL) offers for the creation and interpretation of arrangements for trumpet and live electronics, in order to understand how the trumpeter interacts with the technology generated by the LL and, thus, generate knowledge from the results obtained. The main objective of this work is to present and describe the compositional creation of the work *Densus Bridge* (the second movement of *Trumpet-Loops* piece): for trumpet and live electronics (live looping and effects) with the use of two technologies: loop pedal and guitar processor pedal.

During the research, exploration and experimentation process, we adopted a methodological model called "action research". This is a set of methodologies that succeed each other in a cyclical process – also known as a spiral – which includes three stages: planning, action and reflection, through which improvement is advanced based on previous experiences acquired in the process (Robson 2011).

In turn, the creation and performance process went through three phases. The first was *assimilation*, where we tried to understand how the loop pedal works, how and when to activate the pedals, and how the sound flow happens from the beginning of the sound matrix to the final result in the speakers. This assimilation process took the longest because of insufficient practical experience. Therefore, a period of adaptation was necessary for the use of the loop pedal, as the focus was on the precise coordination of the feet from the first moment.

After a few weeks in this assimilation process, we moved on to the second phase: *exploration*. It was at this stage that we decided to use the effects of the sound modulation processor pedal (pitch, shifter and rotary). However, during the exploratory study – specifically during the performance – the loop pedal effects had some instabilities and delays. To solve this difficulty, we chose to use the processor and the sound effect pedal concomitantly with the loop pedal. With this combination, we were able to solve the problems initially detected in this exploration phase and avoid instability during the performance.

The third and last stage was the **expansion**, in which we sought to explore the interactive character between the performer and the technological tools, loop pedal and sound processing pedal, that is, we search for possibilities of using this object (technological tools), especially the interactions between the performer and the object during the affordance process (Gibson 1979: 134).

It was in this phase that we put into practice the sonic sketches and ideas of musical interaction between the acoustic instrument and the LL technology created during the studies of the two previous processes and the interaction to find a desirable sonic result in the performance. In this work, when citing technological tools, for the loop pedal, we will use the abbreviation LP and for the guitar processor pedal, we will use GPP. As for the structure of this work, it is presented in three parts. In the first part, we will briefly talk about playing the trumpet from the application of electronic devices and the first appearances of the LL in trumpet practices. It is precisely this advent that most interests this work. We will also approach the compositional structure of Densus Bridge, including the creation of the drones and the process of creating the melodic line.

In the second part, which relates to the practical experience, we will discuss two basic points: (1) the experiment itself, and (2) the difficulties/solutions encountered during the process. Finally, in the last part, we will talk about the technical requirements of the work and the configuration we used for the performance of the LP and sound processor pedal, and our reflections and conclusions.

Trumpet and electronic devices

In the 20th century, many avant-garde styles contributed to the emergence of new trumpet repertoire, concepts, and performance techniques. Serial music, *concrete music*, electronic music, and the use of new technologies in musical composition are examples of this. Repertoire for trumpet using electronic music and new technologies had a significant impact on the music of that century (Holmes 2013; Sulpício 2012). Analyses of the trumpet repertoire in the second half of the 20th century shows three phases: consolidation (1950), revolution (1970), and reflection (1990). It is in the period known as revolution (1970) that the production of works for trumpet and electronics began (Muller 2017). Following this same direction, Barth (2011) and Siegel (2017) state that the use of the electronic device in trumpet performance is a new element compared to the traditional repertoire. Siegel (2017) states that the first compositions for trumpet and electronics date from 1965, one such work being *Chaconne* composed for trumpet and electronic sound by composer Henk Badings also mentioned by Barth (2011).

From the bibliographical survey and interviews, we found that few trumpeters use this type of repertoire in their performance, therefore, there is low dissemination. For Barth (2011), this insufficient exposure, combined with the unpopularity of contemporary music, results in a lack of knowledge of such practice. The author claims:

> It is difficult to say with any certainty why this repertoire remains relatively unknown with few publications and recordings. There are undoubtedly several contributing factors, one of which being that it falls within the context of contemporary music which itself tends to be relatively unpopular among performers and audiences alike. (. . .) Finally, it may simply be because it is still a very young genre that has not had much time to establish itself among the standard trumpet repertoires.
>
> *(Barth 2011: 17)*

According to the author Cherry (2009), in electronic manipulation involving a brass instrument, the performer aims to create unique sound effects, generally employing extended techniques in the process, such as the use of half piston or half valve, microtone manipulations and muted tones. In this way, they create distinctive sounds that lend themselves well to electronic processing. For Cherry (2009), the repertoire for trumpet and electronics generally focuses on altering the timbre using pre-recorded tape, digital delay and Midi processing.

The first compositions using real-time electronics called "live electronics" in trumpet performance appeared in the 1970s, where audiotape was used to create an electronic loop or delay effect in real-time (Barth 2011).

With regard to works for trumpet that use a device such as the LP machine and pedal processor in the live looping performance, we didn´t find any with the characteristics of *Densus Bridge* or at least none similar to that work.

Densus bridge and compositional structure

Understanding the sound source controls of the LP processor in the live looping performance

The creation process of *Densus Bridge* was developed collaboratively by two researchers. Initially, we had no experience of creating any musical work using LP and GPP. Up to that moment, our main references were artists who use the LP as another device that follows their performance, along with their pedal sets, as an accessory rather than a tool necessary for creation. As an example, we can mention the performance British artist Edward Christopher Sheeran (better known as Ed Sheeran) in of *Shape of you* (Sheeran 2017).

Through Google and YouTube research, we looked for artists who use or have used LP as a tool for both their creation and their LL performance. A fellow trombonist who was practising at the LoopLab at UA showed us a video on YouTube (Vpro Vrije Geluiden 2015) of the work *Slipstream* for solo trombone and loop station, composed by Florian Magnus Maier. It was from observing that composer's work that we also began to think about other possibilities for the use of LP in LL performance. Unfortunately, we were not successful in our attempt to contact the trombonist who performed the work; However, we did make contact with the composer, who kindly agreed toa short interview by email and answered my questions.

Of the questions we asked Florian Magnus Maier, we highlight three here to specifically support the compositional development of the work *Densus Bridge*:

1) What was the purpose of using the LP in his work?
2) What was the goal when using the LP in *Slipstream*?
3) What were the performance difficulties regarding the use of the LP?

In response to these questions, Maier said:

> 1) Writing a solo piece is always a challenge, and I like when you don't have to limit yourself to creating something that sounds like the rest of the ensemble just didn't show up. So, my departure point is always the music I'd like to hear, not what the available orchestration implies. The work is then to find ways to make sound what I want to hear with whatever's available. Especially for monophonic instruments like solo winds, this can be a big challenge, because I don't want the listener to think "that's pretty cool for a solo piece", I want them to just enjoy the piece without thinking about the fact that there's just one player. In this, the loop station is a great tool for solo artists to expand their possibilities.

It's quite common for singers, guitarists etc from other genres, but in the classical repertoire, it's still somewhat of a novelty.
(Maier 2019: intv.)

2) No, the machine was as new to Jörgen van Rijen[1] as it was to me. What adds to the experience of the piece is that, in contrast to a pre-recorded backing track that is just being played back from a laptop, the public gets to watch the different layers of the piece being built note by note right there and then. This puts a lot of responsibility on the player's shoulders because any mistake he makes will haunt him for the rest of the piece. So, besides the obvious challenge of playing a difficult solo part convincingly on the trombone, the player also has to control the loop station flawlessly, or the piece collapses like a bad soufflé, and the only way to fix it during the performance is to start everything all over again. A bit like doing a tightrope act blindfolded or something. The level of concentration needed from the player is borderline inhuman... but the people who have taken on this challenge are world-class musicians, and hence capable of pulling it off. But I don't want to know how many countless hours they had to put into practising spreading their attention between playing and pressing the right pedals at the right time until it worked as planned. I tip my hat to them!
(Maier 2019: intv.)

3) The loop station gives me the option to create a counterpoint between what the soloist plays and what he played just a few seconds before. Like this, the player creates his/her accompaniment himself - think of percussion, bass lines, chord accompaniments.
(Maier 2019: intv.)

As a result of this interview, we started to consider new ways of using LP in LL performance, making its use obvious, as is done by several artists (Duarte 2016). These possibilities of interaction are called "affordances". This terminology was created by the American psychologist Gibson (1979) to designate the possibilities of interaction between the individual and the environment, that is, to observe the physical properties of a certain instrument or object in terms of its use.

From the perspective of the authors Szokolszky and Read (2018: 9), there is a common relationship between the environment and the agent, contributing to new functions in the observation process. McGrenere and Ho (2000: 1) state that the possibilities of action available in the environment for an individual have always existed, regardless of the individual's ability to perceive these possibilities. From this perspective, here we were not going to create affordancesin LS and GPP to work in LL, but rather identify them in the process of exploration and perception and develop action and learning in the exploratory process.

We started exploring other ways to use the LP, that is, instead of recording phrases or motifs that could be used during LL performance on the tracks, we recorded sounds to create overlapping audio layers (overdubs). These sounds would be our texture sound (drones), the subject of the next topic. In addition to the LP tool, the GPP is used to highlight the sound effects. Therefore, during the *Densus Bridge* performance, the audience will have two sound responses: the real sound of the acoustic environment and the processed sound reproduced by the speakers.

Creation of drones

The use of drones in music dates to medieval Europe and cathedrals such as Notre Dame in Paris, where the huge organs and their pedals were harmonically adjusted to specific tones accompanied by drones or vocal recitatives (Boon 2002: 61). For Hainge (2004), the drone is a continuous note that plays an accompaniment role for other musical events, a characteristic of Russian, Greek and Bulgarian Orthodox Church vocal music, especially common in Hindustani music. Following this same perspective, Courtney (n.d.) states that the drone is an essential part of traditional Indian music, and its function is to provide a firm harmonic basis. In the mid-twentieth century the drone reappeared in the work of La Monte Young's trio for strings (1958), and with the emergence of minimalist music, the drone began to be used more frequently (Votta Junior 2009: 115). Votta Junior (2009) considers the drone to be one of the first aspects of minimalist music, with low notes on which melodic lines are developed. In *Densus Bridge*, the drones are the result of the processor pedal that was programmed to sound an interval of one fifth below the actual sound and aims to create slow sound textures. This process is done as follows: the drones are recorded at the beginning of the work on the LP and used during the performance gradually, being the basis of the melodic line. The sound basis of the drones is the blues scale; the scale was divided into three parts forming four simultaneous sounds of their respective degrees.

In addition to the drones, it was possible to choose which mutes to use during the performance, considering the timbre that each mute has to mix timbres along with the pedal effect. Three types of the muted trumpet were used during this process: (1) Harmon without stein, (2) straight, and (3) cup.

As for the structure and compositional organisation of the drones, the I, V, and VII grades of the scale form a drone group. In the formation of these drones, four successive notes of each grade are grouped to form a cluster. Thus, the drones will have two sound groups: the real sound and the sound coming from the LP and the pedal processor, as seen in the structural organisation of the drones presented in Table 6.1.

TABLE 6.1 Creating drones with the division of the scale

Drone Track 1	Drone Track 2	Drone Track 3
Harmon mute	Straight mute	Cup mutes
I – IIIb – IV – V#	V – VII – I – IIIb	VIIb – I – IIIb – IV
Drones 1 LP e PP	Drones 2 LP e PP	Drones 3 LP e PP
I – IIIb – IV – IV#	V – VIIb – I – IIIb	VIIb – I – IIIb – IV

The drones are recorded on the LP at the beginning of *Densus Bridge*. In bars 2 to 5, the first group of drones are recorded on Track 1; in bars 8 to 11, the drones are recorded on Track 2; and in bars 14 to 17, the third group of drones are recorded on Track 3. At the end of each track, the LP expression pedal is activated to create a decreasing effect. After this diminuendo, the playback on the LP of each drone that was recorded is triggered: Track 1 on bar 20, Track 2 on bar 22, and Track 3 on bar 23. Thus, the three tracks are presented together creating a sound density. After the presentation of the three tracks, Tracks 2 and 3 are stopped, leaving only Track 1 as the harmonic basis for the beginning of the melody line.

After playing the first phrase of the melody line, the drone of Track 2 on the LP is triggered on bar 34, and the third drone of Track 3 is triggered on bar 45. The combination of drones and voice movements brings a sound texture.

Melody line

During the exploratory process of studying the GPP, we sought to find ways of use (affordances) that could be employed during the performance on the LL and that would be distanced from the conventional use of this pedal. We emphasise that for the performance of this work, the processor pedal is one of the necessary components for the creation of the melodic line. To get to this process of creating the melodic line, it was necessary to go through three phases: (1) definition of the sound base, that is, when we chose the minor blues scale, the notes emitted by the performer on the trumpet were just the notes of the blues scale; (2) definition of timbres and intervals of each sound generated by the trumpet and configuration of the intervals and their respective effects; and (3) configuration of the guitar processor pedal to produce two sounds, the first sound with an augmented second interval below, and the second interval with an augmented fourth above. However, in accordance with the movement of the expression pedal, producing these two sounds involves making movements contrary to their respective "directions", that is, when the note processing pedal is activated, the second major interval note below the trumpet sound moves upward, and the augmented quarter note moves downward. Whenever you press the pedal, this movement occurs with these sounds. These two intervals have the real sound emitted by the trumpet as a sound reference. Accordingly, at the time of the performance, three sounds will be heard: the main voice emitted by the instrument and two voices resulting from the guitar processor pedal.

The use of drones along with the melody line starts at bar 27 with the first drone that was recorded on the LP on Track 1, then at bar 33 Track 2 is triggered with the second drone, and the last drone on Track 3 is triggered at bar 45.

Technical requirements for *Densus Bridge*

For this version of *Densus Bridge*, we recommend the use of a LP that provides three separate tracks, in addition to an effects pedal that enables the programming of the described effects. For the performance of the work, the following resources were used: 1) trumpet in B-Flat 2) guitar processor pedal, 3) loop pedal, 4) cardioid microphones, 5) mixer, and 6) amplified speakers (Figure 6.1).

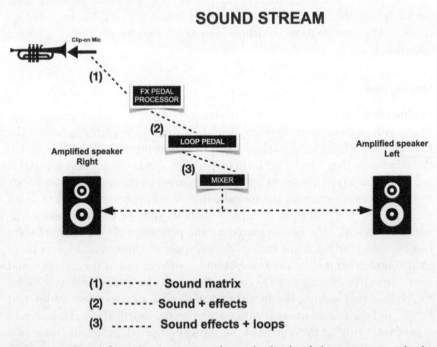

FIGURE 6.1 Sound flow diagram (view from the back of the stage towards the audience)

Image created by the authors.

Loop pedal setup

It is important to emphasise that, for the performance of the complete work, the LP pedal uses three programming banks (patches): one for each movement. This is because each movement uses a certain type of measure and tempo in addition to specific settings to change the equipment's operating schedule. For the performance of the *Densus Bridge* movement, the patch uses the default setting, which is the most common way to use it, as the tracks are synchronised and are of the same size (number of measures). The expression pedal is assigned to control the overall output volume. Thus, the constructed loops are short and are intended to create a "freeze" effect. A technical difficulty encountered at this point concerns an unwanted sound that reveals the "glue" of the loop, a kind of "clip". To hide this effect, it is a good idea to slightly extend the overdubs to camouflage this clipping point.

Pedal processor configuration

Two different patches are configured in two sections: 1) for building drones, and 2) for ground performance. In the first, simpler, case, a pitch shifter or harmonist effect is used to emulate a sound simultaneously with the original, a diminished fifth below. The second, more complex than the first, emulates two sounds simultaneously with the original located, respectively, an augmented second below and a fourth augmented above. This programming acquired a new element that consisted of changing the pitch of these intervals. Each of these intervals changes simultaneously, in opposite movement, going from one point of origin to the other, taking the two intervals as a reference; that is, an interval of a major second below rises to an augmented fourth above; and the interval of an augmented fourth above descends to a major second below (always around the original sound). The control of this change is carried out by the expression pedal, which can go from the very bottom to the top.

Difficulties and technical solutions

Finding other ways to use (affordance) the loop pedal and the guitar processor pedal

In this section, we discuss the difficulties we experienced and the paths we had to follow during this process in order to find viable solutions for the use of LL performance. It is worth emphasising that, in our performance practice, the use of LP or GPP, as well as other types of pedals for sound effects, during performances, represented an unusual and innovative experience.

However, in the creative and compositional process, the three main difficulties we had initially when using the LP and pedal processor were: (1) how to use the pedals of these devices in LL performance in a musical way; (2) how to musically organise all of this sound information in the score; and (3) how to find the most convenient type of microphone for the performance of this work Therefore, in order to solve these challenges during the exploration process in LoopLab, we needed to look at the LP and the processor pedal not only as technological tools, but also as other instruments that needed study and practice.

During the exploratory development of the LP and the pedal processor, we came to understand better, and from a different perspective, the functionality of these devices and found other ways or affordances as to their use. So, after these processes, we started to create the compositional part from the combinations of the LP effects and the expression pedal of the pedal processor, to create a desirable sound in live performance. Overcoming this exploratory study phase, experimentation, and the definition of the sound effects of the pedals, we arrived at the second technical difficulty, which was the organisation of the writing of the composition in the score.

The musical writing was organised in the score as is usual in an orchestral score because we would like it to sound as prescribed, that is, each instrument or voice in each system in the stave, making it possible to view the work in the work. Therefore, we chose to organise each LP track in this way, and it will be available in each system, as in Figure 6.2.

Figure 6.2 shows how the systems are organised. The larger musical figures are the sounds emitted by the trumpet, and the smaller musical figures beside and below them, in the form of x-noteheads, are the sound resulting from LP processing and the pedal processor. The graphics below the melodic

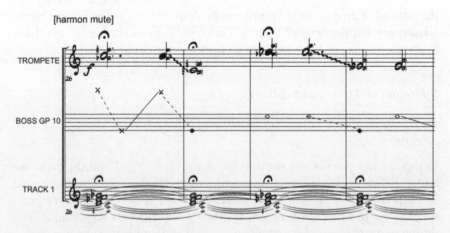

FIGURE 6.2 Organization of musical writing in the score, with each instrument or voice in each slave system

Created by the authors.

line system indicate two movements: (2) normal line fast movement pedal and the processor pedal (1) dotted lines: slow movement of the pedal and the processor pedal. This movement is the opposite of the processing voices. And finally, during the exploratory study of the LP and the pedal processor in the LL performance, specifically in the *Densus Bridge* performance, we used the pedal microphone. With this type of microphone, it was necessary to keep the trumpet bell always directed towards the microphone. However, as the movements of the processor pedal are made with the feet, when moving to use the pedal, it is common for the trumpet bell to move away from the microphone and, therefore, the captured sound is compromised. For a more satisfactory performance of this work, we chose to use a clip-on microphone, as this allows more freedom during the LL performance regardless of the trumpeter´s movement and gesture. Moreover, it perfectly captures the sound of the instrument, as desired by the performer.

Performative analysis

"One way to add a dimension to your playing is to use electronics (...) which may require you to play at certain volumes and with sounds that are not readily available on the trumpet" (Pedersen & Dörig 2014: 29).

"The biggest challenge here, as both performers and composers of the work, was to rethink our musical performance in order to find a desirable sound result whilst performing the work" (Dalagna, Carvalho, & Welch 2020: 10).

In this section, we analyse the first impressions, sensations, and important aspects of the interactions during the performance, since it is a performance using an electronic device, therefore each presentation is unique, and each performer has very specific perceptions during the process.

McNutt (2003) states that musical interaction through electroacoustic music takes on new meaning, and his approach is based on the musical relationships established between individuals and electronic media. This interactivity in electroacoustic music has two variants: (1) from the composer's perspective, the electronic media responds to the performer's action; and (2) from the performer's point of view, the performance itself has what McNutt calls prosthetic elements that can hinder the development of his art such as: pedals, sensors and other new instruments, can be even more invasive than amplification (McNutt 2003: 298). Macnutt argues that the less familiar a technology is to a musician, the more confusing it becomes. Even for big and talented artists. Such difficulties can compromise the success of the performance (McNutt 2003: 299).

In the performance, we sought to put into practice what was developed in the exploratory process from our understanding of the processor pedal and the LP, interacting with new instruments, and seeking to extract from the technological tools new ways of creating music. We should emphasise that during the performance we came up with a sound result that was a little more intense than usual. Previous experience of acoustic concerts contributed to a better understanding of the performance environment, for example, we could consider lighter articulations, when in a room with more reverberation. Issues like this influence musical performance. In the LL performance using technological tools, we are dealing with live sounds, and with that, sometimes, the sound response from the speakers may not be the same as in the preparation process, and we do not have enough time to get to know the room, so what we have here is a contrast between two worlds: the acoustic and the electronic. On this, use McNutt (2003) states:

> Concert music performers, in contrast, are trained to 'play the room', adapting physically in real time to acoustical phenomena. The two are mutually contradictory, particularly if the performer is amplified (which I prefer in electroacoustic music, as it improves the blend and balance of the sound). (. . .) Internal microphones, pickups and cables can drastically alter the weight and balance of an instrument and hamper the performer's movement with a tangle of delicate wires.
>
> The player's physical and sonic identity is significantly altered by the prosthesis of amplification.
>
> Because it is seldom possible to have extensive rehearsal with the audio system and acoustic environment of performance, a strong and trusting relationship between performer and sound engineer is vital.
>
> *(McNutt 2003: 298)*

The performance of the *Densus Bridge* piece requires a lot of concentration, mainly to trigger the LP pedal and drones during the LL performance. As for the effects found in the processor pedal, there was a long process of choice and experimentation in the laboratory because the LL performance brought us exactly the desired result. The use of these technological tools during the performance of *Densus Bridge* created an ethereal atmosphere resulting from the combinations of the trumpet and the mute, together with the processor pedal. The result is the changing sound of the trumpet giving a singular colour and bringing a very specific sound during the performance. This is one of the first objectives of the LL performance in this work: to seek new sonic possibilities for the trumpeter based on these combinations (trumpet and LP).

Conclusions

Raymond (2020), in an interview for *Downbeat* magazine in 2020 about the use of the effect pedal on his instrument, attests that the final objective of the performer in the use of pedals is to develop the flexibility to use them in a way to best serve the music. In our experiment, we realised that using the LP pedals and the pedal processor brought new possibilities for performance. However, the creation of the work *Densus Bridge* was a great challenge, especially for one of us, whose experience focused on classical and popular music, and who had not used any type of technological tool for composition for trumpet, their closest experience being small works and studio productions and performing LL using LP.

We felt very motivated to work in both areas of this research (participating in compositional creation and performing) and experience a bit of these two worlds. It is interesting that, as creators, we always thought of what was best or what could be more useful for artistic performance McNutt (2003: 303) claims that "When performers are fully engaged in the process of creating electroacoustic music, their contributions can be of great value". We emphasise here the support of LoopLab in the processes of research and artistic and compositional creation.

The main objective for Densus Bridge was to demonstrate new ways of using the sound effects of the pedal used together with LP in the LL performance. In this process, we tried to reflect an adequate balance between repetition and contrast. The result of this creative process was divided into three stages: (1) the process of exploring the technological tools in LoopLab that sought to understand the functioning of the LP and the processor pedal and the interactions with the trumpet in the LL performance; (2) the assimilation process important for solving some issues of the inconsistency of the effects of LP sound modulations. For this, the sound modulations effect pedal was used to improve the results; and (3) the amplification related to the process of musical creation and interaction on the LP, such as recording sequences of the intermediate voice layers, bass line and percussion effects with the instrument itself (Duarte 2016).

"It was in the final part of the performance that we consolidated the understanding of the uses of technology, finding a desirable artistic sound result that we would consider innovative for the performance" (Dalagna, Carvalho, & Welch 2020).

Two contributions that we highlight as being particularly innovative in this research: 1) the *Densus Bridge* composition for trumpet using LP and processor pedal is an artistic creation that brings new possibilities for trumpet players to explore live looping in musical presentations; 2) *Densus Bridge* is not a work exclusively for trumpet, it can be performed by other musicians who want to know and play the work using technological tools such as LP and GPP.

The general objective for *Densus Bridge* was the expansion of reflection and the acquisition of compositional techniques in which the focus was the interaction between performance and technology, and the specific objective was to compose a work that would reflect the development and creation of the mastered techniques. As a result, we hope that this will give rise to other work related to this field of research, especially trumpet experiments focused on the use of the LP in LL performance.

Note

1 Jörgen van Rijen, solo trombonist with the Royal Concertgebouw Orchestra, regular member of the Lucerne Festival Orchestra conducted by Claudio Abbado. He teaches at the Conservatory of Amsterdam and is an International Visiting Professor at the Royal Academy of Music, London. He is also a chamber musician in ensembles such as the New Trombone Collective, KCO Koper and Brass United. His special attention is paid to promoting his instrument, developing a new repertoire for the trombone and bringing it to the attention of a wider audience.

References

Barth, Michael Edwin. *Music for Solo Trumpet and Electronics: A Repertoire Study* (Ontario: University of Toronto, 2011). https://tspace.library.utoronto.ca/bitstream/1807/31687/1/Barth_Michael_E_201111_DMA_thesis.pdf.

Boon, Marcus. "The Eternal Drone: Good Vibrations, Ancient to Future." In *Undercurrents: The Hidden Wiring of Modern Music*, edited by Rob Young, 1st ed., 290 (New York/London: The Wire/Continuum International Publishing Group, 2002).

Cherry, Amy K. *Extended Techniques in Trumpet Performance and Pedagogy* (Cincinnati: University of Cincinnati, 2009).

Courtney, David. n.d. *Drones in Indian Music* – Chandrakantha.Com. Chandrakantha: Chandra & David Courtney's Homepage. https://chandrakantha.com/music-and-dance/i-class-music/drones/.

Dalagna, G., Carvalho, S., & Welch, G.F. Desired Artistic Outcomes in Music Performance (1st ed.). (Routledge, 2020). https://doi.org/10.4324/9780429055300.

DeVoto, Mark. "Drone." In *Encyclopedia Britannica*. (2011). https://www.britannica.com/art/drone-music.

Duarte, Alexsander Jorge. "A Arte Do Looping: A Loop Station Como Instrumento de Prática Performativa Musical." *Post-Ip: Revista Do Fórum Internacional de Estudos Em Música e Dança* 3 (2016): 9–20. https://doi.org/10.34624/POSTIP.V3I3.1585.

Dunn, D. "A History of Electronic Music Pioneers", in D. Dunn, *Eigenwelt der Apparate-Welt: Pionieers of Electronic Art* (New Mexico: Ars Electronica, 1992), 21–62.

Gatti, B. A. "Pesquisa, Educação e Pós-Modernidade: confrontos e dilemas. " Cadernos de Pesquisa (Fundação Carlos Chagas), São Paulo 35 (2005): 595–608.

Gibson, James J. *The Ecological Approach to Visual Perception*. 1st ed. (Boston, MA: Houghton Mifflin, 1979).

Griffiths, Paul. *A Música Moderna* (Rio de Janeiro: Jorge Zahar, 1987).

Hainge, Greg. "The Sound of Time Is Not Tick Tock: The Loop as a Direct Image of Time in Noto s Endless Loop Edition (2) and the Drone Music of Phill Niblock." *An Electronic Journal for Visual Culture (IVC)*, no. 08: The Loop as a Temporal Form (2004) (October). http://www.rochester.edu/in_visible_culture/Issue_8/hainge.html.

Hobsbawm, Eric J. *Era Dos Extremos: O Breve Século XX (1914–1991)*. 2nd ed. (São Paulo: Companhia das Letras, 1995).

Holmes, Aaron David. *The Evolution of Instrument Design and Its Influence on Trumpet Repertoire* (Maryland: University of Maryland, 2013). https://drum.lib.umd.edu/handle/1903/14401?show=full.

Kerr, Dorotéa. 2012. "A Música No Século XX." *Objetos Educacionais Unesp*. São Paulo.

Mcgrenere, Joanna, & Wayne Ho. "Affordances: Clarifying and Evolving a Concept." *Proceedings of Graphics Interface 2000*, 2000.

McNutt, Elizabeth. "Performing Electroacoustic Music: A Wider View of Interactivity." *Organised Sound* 8, no. 3 (2003): 297–304. https://doi.org/10.1017/S135577180300027X.

Meyer, Felix, Carol Oja, Wolfgang Rathert, & Anne Shreffler. "Crosscurrents: American and European Music in Interaction, 1900–2000." *American Music* 32, no. 3 (2014): 366–370. https://doi.org/10.5406/AMERICANMUSIC.32.3.0366.

Muller, Aaron Douglas. *Consolidation, Revolution and Reflection: Music for Trumpet from Three Decades 1950s, 1970s and 1990s* (Maryland: University of Maryland, 2017). https://drum.lib.umd.edu/handle/1903/19915.

Pedersen, Craig, & Ueli Dörig. *Trumpet Sound Effects* (Boston, MA: Berklee Press, 2014).

Raymond, John. "Pro Session: Using Effects Pedals on Your Horn." *Downbeat*, 2020.

Robson, Colin. *Real World Research: A Resource for Social Scientists and Practitioner-Researchers*. 4th ed. (Padstow: John Wiley & Sons, Ltd., 2011).

Siegel, Steven. "A Performer's Guide to Works for Trumpet and Synthesizer by Meg Bowles" Theses *and Dissertations—Music.* 85 (Kentucky: University of Kentucky, 2017). https://doi.org/siegel87@gmail.com Digital Object Identifier: https://doi.org/10.13023/ETD.2017.156.

Sulpício, Carlos Afonso. *Transformação e Formação Da Técnica Do Trompete: De Monteverdi a Stockhaunsen* (São Paulo: Universidade Estadual Paulista, 2012). https://repositorio.unesp.br/handle/11449/103395.

Szokolszky, Agnes, & Catherine Read. " Developmental ecological psychology and a coalition of ecological–relational developmental approaches. " *Ecological Psychology* 30, no. 1 (2018): 6–38. https://doi.org/10.1080/10407413.2018.1410409.

Votta Junior, Alfredo. *A Musica Tintinabular de Arvo Part* (Campinas: Universidade Estadual de Campinas, 2009). https://doi.org/10.47749/T/UNICAMP.2009.473447.

Vpro Vrije Geluiden. "Jörgen van Rijen – Slipstream for Trombone Solo and Loop Station (Live @Bimhuis Amsterdam) ". Youtube. November 30, 2015. https://www.youtube.com/watch?v=J2dqO5PQGTE.

Interview

Maier, F.M. Interviewed by Elielson Gomes. May 14, 2019. By email.

7
PERHAPS THE LOOP STATION IS NOT THE POINT

José Valente

The singularities of a composer

Some years ago, pianist and composer António Pinho Vargas gave a few masterclasses at the Escola Superior de Música de Lisboa, in which he introduced several key concepts to understanding different creative attitudes. Among the shared stories and examples, he organised the compositional process in the following fashion: the creator has a crucial *idea*, a concept, which feeds the *structures* and *musical gestures* that will define a future composition. This terminology isn't very far from Cage's initial configuration, *material, structure* and *method*, as explained in *Silence*.

Any composer has their own dilemmas and routines regarding a new work. And, in many cases, the particularities of each *structure, method* or *gesture* become so embodied within the composer's vision, that they might find it difficult, unless specifically asked to discriminate such elements, to point out which *gestures* influence the *structure*, and vice versa, which one had the edge that encouraged the other. But almost every decent creator will be able to explain the concept, the intention, the core motivator for a new piece of music. And, perhaps after many attempts at conveying their own personal musical flavour, *gesture, structure* and *method*, grow to be so intertwined with each other that, eventually, they turn into a single motion: a kind of musical DNA, a musical vocabulary and language that describes a composer, independently of form or purpose; independently of musical genre and environment. Perhaps, when the composer arrives at such a fine level of expression of their own identity, they end up always creating within a dynamic that is apparently free of *gesture, method, structure*, because such constituents are internalised and delivered in each piece almost automatically.

DOI: 10.4324/9781003154082-9

Improvisation as research: space-time *continuum*

In my experience one can find many precious sound options during improvisational exercises or dialogues. One can invest a period of time just to improvising or enjoying improvisational games, record the improvisational experience, analyse it, and select whatever might be appropriate for a future work (if the composer doesn't play an instrument, they can conduct the improvisation or suggest challenges for the other players). If the composer has already the above-mentioned capacity, let's call it a personal vocabulary, then most of their preferred *gestures* and *structures* will, one way or another, manifest themselves. However, because the situation is unpredictable, especially if the improvisation happens within a musical conversation, for example, their perspective can shift and therefore be seduced by new opportunities and desires.

This practice has another essential aspect that is very attractive: it stimulates the decisive connection between the musician and the *space-time continuum*, a perception that is fundamental for any kind of performative activity (Oliveros 2005). One dangerous, yet usual, argument around this possibility is that in order for you to evolve within this *continuum*, you must dedicate your entire creative energy to only the improvisational hypotheses. However, such a statement has one disadvantage: the improviser might never distance themselves enough from the performative action and, consequently, not be able to analytically criticise the standard of the played vocabulary, or the pertinence of the improvised atmosphere and phrases, entering a loop of constant experimentation and losing track of other, more concrete creative goals.

Considering this process where improvisation is a medium through which to search for alternatives, it is the compromise between the activity experimented with and its analytical feedback that can help one find new solutions for whatever issues the piece or music in general has raised. In several of my works, the performer receives a set of improvisational instructions that allows them to play a certain passage while being aware of the *space-time continuum*. In several of my works, where I am the performer, this possibility even goes further, expanding itself to my entire vocabulary, that I've mastered through the years and throughout my original music.

The *weight of sound*

Any sound can be music, it's all a question of why, how, and when. The purpose behind a sound or the *weight of a sound* is what, one way or another, defines if the sound is or isn't music (Zappa 2012). I believe that every chosen sound has a past, a present and a future. Meaning that the moment the performer plays a sound, it is impossible to overestimate the sound value.

The sound that originated from a context immediately has a past. Then, upon being heard by the listener, the sound initiates a relationship with the listener, based on the memory that the listener has of the sound and its context, and what they imagine the sound will be (Barenboim 2009; McFerrin 2010). Eventually the sound moves forward being both present in the space while drawing a path towards its future. This notion can be extremely helpful for gluing a piece together. It doesn't mean that a composer must work with endless imitative techniques, abusing the flexibility of the invented musical motifs and dragging the listener into a hardcore baroque revivalism. It only ensures that no sound is wasted. That every chosen sound means something. Every sound matters.

In my first album "Os Pássaros estão estragados",[1] the melody played solo and ad libitum in "Eduardo" (which is actually the name of the melody as well, merely for metaphoric reasons) occurs again in "A Voz entre as Cinzas",[2] but this time played an octave higher and accompanied by three violas with rhythmic ostinatos in thirds, resembling the sound of moving wings. The second appearance of this melody, surrounded by a different orchestration and background, offers a conceptual attachment that is sustained throughout the album and gives "Eduardo", as other musical moments, a very significant status within the album's personality.

In my second album "Serpente Infinita",[3] an almost improvised melody played on the viola is presented to the listener in the first track, surrounded by a multitude of dark, ambiguous sounds executed with extended techniques on the viola, with percussion and Constel.[4] This almost improvised melody is delineated by a few elected notes and the way that these notes should be played (*structure* and *method*). The musician must be aware of the encircling ambience as well as the textural potential that each note has (*space-time continuum*), in order to decide the duration of every note. The long notes are to be explored: one can drive the bow in several directions (both horizontally and vertically) and enjoy the various effects that slowly emerge with such action.[5]

These selected notes were then applied in diverse ways through the work/album. They emerged either upfront, serving as a reference or a memory of a feeling suggested to the listener in the first track; or in the background, as a harmonic essence for other melodies, as a bass line for an improvised solo, etc. At the same time, these opening notes also state an intervallic atmosphere that characterises the entire work/album.

In this album opener, a single chord is strummed on the viola. This diminished chord, built from a triton, a minor sixth and a minor third (four notes: e, b flat, g flat and a) functions as a leitmotif as well as a question mark, providing the musician with a musical statement that can respond to certain propositions delivered by spoken word, as well as confront the various realities that are suggested to the listener throughout the entire work.

These are four examples of the aforementioned idea of sound and its past, present and future.

The limitations of a loop station

Art has tremendous potential for elevating singularity, beauty, of founding the unexpected (Calvino 2010). It can take the spectator to a point of transcendence that helps them forget the daily routines, the common life, and achieve an emancipatory level of consciousness (Nuno Teotónio Pereira 2014; Camus 2016).

It is the aim of the composer to invent musical circumstances that motivate subjective thoughts and emotions in the listener. The principal question isn't so much: what style of music is or isn't sophisticated enough to implicate the listener in this stated goal? Nowadays, the question should be more: how can I, the composer, create a musical expression, regardless of style or environment and whatever orientations guide the process, that can provoke the listener to confront their reality, to feel something unpredictable?

The *space-time continuum*, together with the perception of the *weight of a sound* and the freedom to choose any genre of music or musical practice, gives the composer the full resources to obtain new stages of communication and subjectivity. Therefore, the issue to be discussed and critically analysed isn't that a musician uses a looper or any other known technological expansions of the instrument in their work and performance. Quite the opposite. The important conversation has to be around the motives and consequences inherent in the choices that composers and performers make while using such devices.

There is risk in attempting to accomplish some sort of artistic testimony and knowledge while looking exclusively to the instruments that are used. Technology is just one utensil that the musician has at their disposal. Therefore, to have the thinking spotlight on the loop pedal or other electronic gadgets is reducing the incalculable number of nuances that form part of any creative process to a very one-sided topic. Furthermore, in the specific case of the loop pedal, the limitations of the machine make it quite restrictive when facing more ambitious and advanced musical shapes and intents.

There are, around the world, live looping competitions where musicians produce tracks using the various tricks and techniques on their instruments. In most of these competitions the target is to record an accurate groove in the shortest time possible. Usually, it is expected that each contestant builds a series of loops that, together, resemble a rhythm section (bass, drums, and harmony riff), above which the musician can either introduce a melody or improvise. Usually, what is evaluated in such competitions is the accuracy of the recorded loops and the time that the musician needed to construct the groove, loop after loop. So, if a musician finds something suited to these expectations really fast, there is a higher chance of triumph. In a way the loop pedal was invented for this premise. The simple fact that each loop bank has a fixed duration, determined by a precise pulse (that can be muted or unmuted) induces the musician to create or replicate something quite square and banal.

Some artists are, of course, excited by these possibilities. The opportunity to replace bass, drums and rhythm guitar, while playing a single guitar, for example, gives the illusion that the instrument is, in fact, an expanded instrument. And, yes, with the addition of some other pedals (like an octave pedal or a harmoniser pedal, among other options) the instrument does become augmented and permits the musician to explore more frequencies than usual.

Exploration and expansion of the instrument

One can also achieve similar purposes playing a singular acoustic melodic instrument, like the human voice or the viola. Either through the exploration of extended techniques or other natural resources that each instrument has. Or by inventing techniques that unlock other musical roles for the instrument, building its vocabulary and placement in music when compared to more traditional customs.

A good starting point and standard for researching how to spread a melodic instrument beyond its regular, more conventional features is to transcribe and be attentive to the peculiar approach of singer Bobby McFerrin. Or, in a more abstract direction, bassists like Mark Dresser, Barry Guy and Joëlle Léandre, and violinists such as Carlos "Zíngaro", who are specialists in free improvisation and extended techniques; with a more jazz/bluegrass vibe, violinists like Christian Howes, Billy Contreras, or cellist Rushad Eggleston, who are known for playing versions of famous tunes and their original tunes, executing both groove, harmony and melody at the same time on their acoustic instruments; and Peter Evans on trumpet or Evan Parker on saxophone, both of whom create four-voice improvised conversations by using the circular breathing technique. Also, in musical compositions such as Patricia Kopatchinskaja's Cadenza on Ligeti's Violin Concerto, the beginning of Jorg Widmann's Viola Concerto, my piece "Passaporte" for solo viola, the Sequenzas by Luciano Berio for many solo instruments, Garth Knox's "Fuga libre" or "Quartet for one" for solo viola, where all of the aforementioned skills are applied as well. Increasing an instrument with electronic devices can be an interesting proposition but it shouldn't be the driving force behind an artistic personal voice. A composer desires the accomplishment of a musical path and distinction that goes beyond the instrument or the instruments employed. Certainly, there are certain degrees of success and musical careers that grow on more superficial assessments of appreciation, but I believe such directions to be short-sighted.

Many students or aspiring musicians have asked me, after my concerts or during my masterclasses, about the usage of the looper, and I always answer with the same principal: be able to do something special, positive and unique with your instrument before starting to explore the machine; because, at the end of the day, it is that specific touch, that personal input that is going to make a difference and separate your music from everything else that is out there, and not necessarily the fact that you use such tools in your set. Unfortunately, many musicians get stuck on the simplicity of the looper and spend most of their time replicating old musical formulas that are easily produced with such technologies.

Practising with a loop station; subverting a loop station

I started using the looper to learn chord progressions over which I wanted to improvise. Basically, I would record the fundamental, the third and seventh of each chord and then play the respective scales, arpeggios, voice leading, transposed motifs etc. With this practice, together with other exercises, I learned jazz vocabulary and harmony on my viola. I write purposely on my viola because there is an immense difference between understanding harmony from a theoretical point of view, or knowing harmony through your fingers, with the instrument.

Years later I was commissioned to compose a piece influenced by Italo Calvino's "Invisible Cities". Having already been informed that it would be impossible to hire an ensemble, due to the commission's lack of resources, I adapted my initial compositional plans towards a solo option. I wanted, for example, to write a neoclassical opening for a quartet that would expose the ostensible perfectness of an imaginary city. A glorious façade, built with intricate imitative games between the violas, that falls apart abruptly (second movement), giving place to a decadent, busy, noisy city: a hazy musical environment made with field recordings of New York's subway and confused crowd, together with the dissonant glissandos of the violas and a hidden rap by 2Pac.[6]

Until my take on "Cidades Invisíveis",[7] I had composed several short pieces/exercises with the looper, which were part of the initial steps of my extensive solo repertoire comprised of both acoustic as well as amplified works. In these first pieces I not only managed to apply the abovementioned capacities of the loop pedal (recording several layers within a fixed metric), but I also found the set that, later on, permitted me to deconstruct or go beyond these predetermined characteristics of the machine.

In my electronic/pedal set, I connect two loop pedals. This is relevant because it is quite a simple solution to a specific problem forced by the limitations of the looper: how to avoid a long, tedious performative process of recording layer after layer.

With this setup I'm able, for example, to record one bar of a short ostinato that is a core element of a bigger form, and then, without breaking the momentum, record the other components that define the desired bigger form. In a way, this setup offers a shortcut to what can become a very monotonous episode in a live show.

Perhaps such a possibility doesn't completely solve the main issue imposed by any loop pedal: the tendency to invent tunes with repetitive patterns, smashed inside a closed timeline.

The process behind "Cidades Invisíveis" exposes a relevant reason why I ended up utilising the loop pedal. Considering the precarious environment of the art scene in Portugal, especially of contemporary art (financially, as well as knowledge and courage-wise), the opportunity to show my music as a solo act relieved me of a lot of other logistical requirements vital to the subsistence of a bigger ensemble, and made my project and talent much more accessible. It was extremely important for me, when I returned after ten years abroad, to be able to live off my musical activity, instead of other means. Consequently, I had to find, early on, an option that would facilitate my recitals at any kind of venue.

In 2012, I spent five weeks at a ranch in California, having been invited to be a resident artist at the Djerassi Residency Artists Program. Being free, during this time, of almost all my usual daily duties, necessary for building a consistent agenda of concerts and composing commissions, I embraced the raw nature that surrounded me, and I studied, quietly, the way the other residents behaved in this small community of artists. I took many notes and quotes of several reactions and arguments of my colleagues, that were apparently harmless but contained deep levels of selfishness and envy. The witnessed invasion within human relationships, together with the invasion that the ranch and its buildings represented inside an almost virgin territory, gave me the concept, the *idea*, for one of my most significant solo pieces: "Invasão".

It was in "Invasão"[8] that I started to develop a number of *gestures* and *structures* that, eventually, became significant and dear to my artistic path. Although there was already a recognisable vocabulary and discourse in my music, this work established two essential aspects of my way of doing things: the exploitation of spoken word as an integral factor of the music, and finding out how to subvert the defining borders of a looper.

At its live premiere, "Invasão"[9] opened up with a distant, high-pitched sound composed of two dissonant violas playing an ostinato rhythm, retrieved out of transcriptions I did from the singing of the crickets that lived around me at the time. Besides the ostinato, brief and fast melodic motifs started slowly to emerge on other violas, conversing amongst themselves. How could I achieve these dialogues with only my viola and the loop pedal? By creating something that was very acute and clear (the sparse motifs + the ostinato) to be recorded in the looper within a very long timeframe, without obvious layers. By building a series of loops inside this long abstract form, the perception of time would disappear to the ears of the listener, transforming what seemed to be a solo performance into a chamber music moment.

In "Invasão" I've also dealt with another issue intrinsic to the loop pedal. It is something that is often seen in live performances with loopers: the tendency to deliver a step-by-step construction of a tune, a riff, etc. To avoid the monotony that usually involves such a process, I started to insert pre-recorded tracks into the looper, to activate when appropriate, first in "Invasão" and subsequently in other shows and works. This doesn't mean that I never benefit from the repetitive nature of a looper, or that I never expose the trick behind the gadget; it just isn't a primary concern.

But the highlight of "Invasão" isn't the accomplishment of generating a conversation among various imaginary violas. The combination of diverse musical approaches, from contemplative melodies to marches (a pre-recorded viola trio); to the improvised interactions with field recordings from the movie "Alto do Minho";[10] to the satirical lyrics spoken by a deep, rough male voice; to the mixture of musical vocabularies and genres, composed and organised to announce a subjective yet strong impression of the invented *concept*, through well-designed *structures* and *musical gestures*: this was what made the work successful. This is the characteristic that best describes my vision on a strictly technical level of analysis. I absorb and accept any style of music within my realm of motifs and forms. This infinite bank of information is, for me, the absolute musical freedom. It is within musical freedom that I search for ideas, that I am challenged by the unexpected. I am more concerned about my discourse, than any category or compositional practice, because it is the discourse that feeds my artistic authenticity.

In 2012 I was still trying, while crafting "Invasão" for example, to adhere to the conditions imposed by the *material* that was available, in this case the viola plus electronics set-up; to the practical concern of performing the work in a live situation.

However, with the studio experience, I learned that there is no advantage, for the sound attributes of a viola, to record while being connected to devices such as a loop pedal. Therefore, if one wants to record an indeterminate number of loops with a viola, one has, in order to achieve a more precise and round sound quality, to record each looped voice one by one in full, or to record shorter sections and hope that subsequent editing is accurate. Having in mind the amount of time required to capture all the necessary loops, as well as the modus operandi itself, I realised that it would be much more interesting and rewarding, artistically speaking, to assume an orchestration for several polyphonic violas instead of trying to recreate the more inert looping, and thus to develop music for viola ensembles to be recorded by one person.

"Serpente Infinita"

This mindset completely shifted the working strategy. In "Serpente Infinita"[11] the orchestration enjoyed viola quartets, spoken word, punk-rock drum grooves, several synthesisers, several percussion instruments, Constel, and more. All recorded by me, João Rocha (on drums) and Marta Bernardes (spoken word). In "Circulando"[12] the instruments used were pianos, marimbas, organ, a large number of traditional percussion instruments from Portugal and various violas, all recorded by me.

The looper was completely out of the picture during the creative process of both these pieces and many others.

In 2015, I started to reflect upon the subtle oppressive mechanisms that continue to deter Western society from a more humanitarian emancipation. In summary, a series of political decisions together with the promotion of frenetic consuming habits, have been offering and rousing a lifestyle that is oriented towards the hyper-value of material possessions, impeding attitudes that encourage the spiritual and intellectual development of humans. We face a conjuncture where the status quo, being typically a conservative standpoint, is drawn and defended by mediocracy (Rieman 2011).

From this scrutiny and subsequent hypothesis, the "Os Pássaros estão estragados" album was born. However, two years later, and being confronted with the obvious fact that the investigated reality mutated into something much more aggressive, I was inevitably thrown back to the subject and ended up composing a second musical step with a similar motif. The result of this effort was my second album: "Serpente Infinita".

> It happens that the stage sets collapse. Rising, streetcar, four hours in the office or the factory, meal, streetcar, four hours of work, meal, sleep, and Monday Tuesday Wednesday Thursday Friday and Saturday according to the same rhythm – this path is easily followed most of the time. But one day the "why" arises and everything begins in that weariness tinged with amazement. "Begins – this is important. Weariness comes at the end of the acts of a mechanical life, but at the same time it inaugurates the impulse of consciousness. It awakens consciousness and provokes what follows. What follows is the gradual return into the chain or it is the

definitive awakening. At the end of the awakening comes, in time, the consequence: suicide or recovery.

(Camus 2016: 23)

The denouncement of social circumstances in "Os Pássaros estão estragados" isn't a simple criticism of politics and its deeds. We live in a subjugated regime that initiated its tyranny the moment universities excluded themselves from their active and permanent responsibility to determine a demanding standard of knowledge. This passive role has allowed an increase in ignorant protagonists, not only inside the interior dynamics of universities, but also inside the several circles of decision-making that affect the world's function (Foucault 1971).

In "Serpente Infinita", regular life is interpreted as a cause for the apathy and banality, for the agonised, trapped experience that humans go through during the course of every day. The normative routine is a painkiller to relieve the intrinsic anguish that the realisation and notion of the absurdity of life provokes. At the same time, it is an imprisonment of a free conscience and, therefore, of a sensitive, explorative mindset that appreciates the unpredictable aspects of life. "Serpente Infinita" deals with the suffocation enforced by the day-to-day routines that are repeated by the majority of humans, swallowing everything else: repressing the exceptional, the alternative landscapes, into a boring, hypocritical existence.

This was the philosophical core that influenced the origin of the main *idea* and then established significant metaphors that later supported not only the concept, but also the *structure* of the work/album.

It is relevant to mention that around the time I was thinking about these issues, I was invited to do a residency at Musibéria. Musibéria and its record label, Respirar de Ouvido, managed the production and publishing of "Serpente Infinita". The creative process was positively influenced by this logistical context. I arrived at the recording studio with a finished score of the piece. In the first meeting with the sound engineer, I explained the composition step-by-step, having the score as a reference. But, after trying out Constel (an instrument that is in storage at Musibéria) and realising that its sound should fit in this composition, the initial plan started to fall apart. During the first five days of recording, I kept on rejecting the original score. When playing the improvised solo on "Respiração Interrompida",[13] I finally felt that we were moving in the right direction. If I wasn't given the time by Musibéria to deal with such important last-minute indecisions, "Serpente Infinita" might never have become the rewarding album that it turned out to be.

While "Os Pássaros estão estragados" has a conclusive movement that offers some light against a dark situation, "Serpente Infinita" doesn't pretend to be resolving anything. There isn't any solution for the difficult, distressing feeling, bred by the previously-explained imprisonment. "Serpente Infinita" only faces the suicidal consequence of Camus's "*end of the awakening*".

Most of the musical *gestures* are a direct outcome of this feeling and metaphor:

- The heavy metal riffs and rhythms that represent a scream of discontent and revolt. These riffs are also an opportunity, for the listener, to rest from the more vague and frustrated atmospheres. These heavy metal references were created following the transcription of numerous riffs from Tarantula, Children of Bodom, Slayer, System of a Down, Opeth, Korn, among many others.
- The previously-described strummed chord, that often appears in the work.
- The above-mentioned melody.
- The usage of many extended techniques on the viola or on the percussion instruments or Constel. There was research for these sounds and, once founded, they were the essence behind the windy, scary ambiences that surround the album.
- A direct reference to Shostakovich, and specifically the third movement of his Eighth String Quartet, which was a big influence for the melodic material in "Mastigando" and "Todos os dias".[14] This string quartet was dedicated to all victims of fascism.
- The walking ostinato made with uncountable and intermittent, yet metric and straight, quarter notes: always using the same note, the note is repeated a few times before jumping an octave higher for one single note, coming back to the initial note. The jump is irregular; this walking motif is another musical indication of the idea of routine and monotony.
- The artificial sound of the synthesisers, inspired by the electronic music of Oren Ambarchi, Jonathan Uliel Saldanha, and Bob Ostertag; the album "Dark Territory" by Dave Douglas; and by Gabriel Pinto's solo in the tune "Logos" in Hugo Carvalhais's album "Grand Valis".
- The persistent company of long notes that change texture while being played, as a never-ending serpent that continues to viscously and slowly move and feed from day-to-day habits.

Another fundamental aspect to consider regarding my compositional process is the frequent practice of a *composed improvisation* or an *improvised composition*. By understanding the relevance of the *space-time continuum* during a performance, together with the desire of achieving a distinguished personal vocabulary, I realised that it was imperative to establish a deep bond between any composed material and any immediate improvised action. The principal is to accomplish such a clear level of expression that it becomes hard for the listener to separate what is composed from what is improvised.

This mindset is evident in how the initial viola cadenza of "Relações entre indívíduos"[15] (from "Serpente Infinita") was made. Before arriving at the studio, I invented a, supposedly, intense duo moment where the viola would introduce certain heavy metal riffs, and then do variations of these riffs accompanied by the drums. But, after recording what was in the score, I realised that the intentional question-and-answer trade between the viola and the viola plus drums duo was redundant. It was only a basic demonstration of a theory. Proof that the composer, me, had researched and knew the core particularities that frame most of the heavy metal main- and subgenres.

The answer to this problem ended up being an improvised solo constructed from most of the created musical motifs of the section. I improvised a set of takes and then selected the ones I considered the best. Thus, a viola cadenza was created.

Considering this umbilical relationship between composition and improvisation, commissions like "Serpente Infinita" are a great opportunity for me to develop a broader *method*, besides the philosophical concept that nourishes the *idea*. A *method* that also focuses on growing musical material that was inaccessible to me until the starting point of a new piece. A *method* that grants me new, constant information for my improvement as an improviser and violist.

For example, in "Serpente Infinita" the faster melodies are filled with tritons, perfect fourths, and chromatic approaches. As these were intervals with which I had trouble improvising when playing faster passages or tempos, I purposely incorporated these complex intervals for my instrument into my melodic *material* and, later on, melodic *gestures*. There was an unknown terrain that I wanted to explore and learn to a higher level of proficiency: to be able to play such complicated runs while improvising. Through the practice of my new piece, I would also conquer a new set of skills that would help me be a more authentic musician and, also, a virtuoso on my instrument. Virtuosity isn't necessarily evaluated by the speed or by the energy that one performs a certain passage. It is much more about the quality that one reveals when handling a certain technique or vocabulary. Playing a scratched, noisy sound can be a demonstration of virtuosity. It all depends on the *weight of the sound*; on the intention and the ability to execute that intention clearly.

There were therefore, two main concepts behind the *idea* and *method* of "Serpente Infinita":

- To confront the listener with their own quotidian existence, through the suicidal framework that the notion of the absurdity of life can instil in anyone, the moment conscience awakens.
- To build new techniques and to develop newer, exciting language, for my viola playing.

While I was composing this album, I discovered the amazing, contemporary poetry of Ana Hatherly. Ana Hatherly is one of the key figures in Portuguese contemporary visual arts and literature, known for expanding the borders of painting with poetry, and vice versa. For instance, an entire collection of her work is dedicated to the usage of words drawn as the trace and silhouette of pictures. In poetry, her contribution is of historical importance: she, along with other poets like E.M. de Melo e Castro, cultivated an extensive catalogue of experimental and abstract poetry. Enjoying simple exercises such as the syllabic deconstruction of words, she invented a completely different perspective for poetry writing, broadening the canvas to a huge number of original options.

I was amazed by the daring, audacity and rigor of her lyrics. Eventually, I chose four *tisanas*[16] from her "463 Tisanas" to join the universe I was creating in "Serpente Infinita". The *tisanas* are very short announcements, stories (one or two paragraphs, not more) that reveal reflections, imaginary worlds, situations and much more.

One of the selected *tisanas* was number 68. The first sentence of this lyric is *"Era um vez uma serpente infinita"*[17] and that's how I found the name "Serpente Infinita" for the work/album.

> *Era uma vez uma serpente infinita,*
> *como era infinita não havia maneira de*
> *se saber onde estava a sua cabeça. de*
> *cada vez que se lhe tirava uma vértebra*
> *não fazia falta nenhuma. podia-se mesmo*
> *parti-la deslocá-la emendá-la. ficava sem-*
> *pre infinita. quem quisesse levar-lhe um*
> *bocado para casa podia pô-lo na parede e*
> *contemplar um fragmento da serpente in-*
> *finita.*[18]

The *tisanas* provided four marks – four arrival spots – that helped me define the final form of the piece, as well as an assembly of images and thoughts that asserted the metaphor that I intended to raise through this music.

Tisana 68 [*"Once up on a time there was an infinite serpent (. . .)"*] introduces the listener to the conceptual proposition that sustains the album. *Tisana 55* [*"The relationships between individuals are so frightening (. . .)*[19]] exposes the listener to my own conflict when facing the slow invasion of the daily routines that surround me. *Tisana 81* [*"Once upon a time there was spaghetti in blood. Every day, at one o'clock in the afternoon (. . .)*][20] embellishes an extremely scary environment that is perpetuated by the employees of a Ministry, a conclusive ambience for the work/album that ends with *Tisana 69* [*"History is infinite. We can intercept it/in any point, once upon a time there/was a city where its inhabitants knew/so much about the human suffering that when/they woke up, they would lay down immediately."]*[21]

Besides the *tisanas*, there is also one poem written by me, incorporated in the first movement (the first three tracks of the album) of this piece; an ironic, denouncing, poem that manifests the oppressive lifestyle that comes from a repetitive routine.

I wasn't, at any moment of this process, concerned with the use of a looper. The loop pedal was never part of my creative dilemmas, neither was it used during the recording of "Serpente Infinita".

However, this album was performed live with three kinds of instrumentation:

- As a solo act, where I mostly used the loop pedal to shoot pre-recorded synthesisers and spoken word. On one or other occasion, I applied the looper to make abstract sections with extended techniques, or to make groovy ostinatos, over which I improvised solos.
- As a duo act, with pretty much the same configuration as the solo act but with live spoken word.
- As a trio: one musician playing drums, percussion, vibraphone and electronics, another playing synthesisers and electronics; and me on viola and electronics.

None of my live shows were an exact representation of the recorded music of the album. Although the work was the same, my objective with live concerts was, and is, to enjoy the composed music as a background for creating other artistic moments and experiences.

"Serpente Infinita" was, for reasons that I don't really understand, announced by music critics as a solo album, despite it having so many textures and instruments throughout its musical story. However, such a categorisation was extremely pertinent to my decisions regarding the future. In the same way that it is fundamental to judge the state of one's ability each time there is a commitment to create a piece, so too can one find other ideas to pay attention to, as well as other technical challenges; and it is also essential when evaluating the state of the art (in this case, an individual state of the art) to analyse the musical path already walked.

Bearing in mind the processes and the guidelines behind works like "Os Pássaros estão estragados" and "Serpente Infinita", it became imperative for me to aim for motivations different to those that steered both these works/albums.

What's next, what's on

I decided that my next pieces should manifest a shift towards an even more ambitious musical reality. Being aware that my reputation as a solo musician was established, I finally started to be in a position to organise projects that required bigger logistical demands. The third album of this trilogy, following "Os Pássaros estão estragados" and "Serpente Infinita", was "Trégua",[22] an unprecedented alignment of new pieces for viola and a marching band, the Orquestra Filarmónica Gafanhense, conducted by Henrique Portovedo, that focuses on the feeling of re-establishment, of joy and fun as a way of subversion.

In the meantime, I composed a work for violas, organ, piano, percussion, choir and spoken word, created in collaboration with writer Gonçalo M. Tavares, commissioned by MAAT (Museu de Arte, Arquitectura e Tecnologia). My last album was "Águas paradas não movem moinhos", a homage to the music of José Mário Branco, arranged by me for my newest ensemble, a viola sextet. Recently, I was commissioned to compose a series of songs as well as a major orchestral piece, through an inspiring connection with several poets and five different ethnicity groups that reside in Porto.

Some of my more recent pieces like "Passaporte" or "Trégua" already demonstrate a more optimistic approach in comparison to the standpoint that originated "Serpente Infinita". For "Trégua", I researched laughter and its characteristics on a social and spiritual level, and I found some very interesting points. Notions such as: when on stage, the comic can be a pretend puppet of the audience, by relating adventures that are familiar to the life of the spectator, to then, with the punch line, reveal that the comic was never the puppet but the puppeteer (Bergson 2009); or basic comedy techniques such as: being funny by opposition; being funny by imitation; being funny by augmenting a situation to its exaggeration; being funny by changing the context of a situation; or being funny by repeating a situation many times (Araújo Pereira 2019).

Comedy has huge potential for the unpredictable. So, in a way, comedy is always a form of subversion, of defying the system. But that is a subject for another chapter.

Notes

1 Os Pássaros estão estragados (*The birds are broken*), 2015, JACC Records. Listen here: https://open.spotify.com/album/6GREF9jKza8FPcPtW5DkfL?si=V1l3O3N xQiOmo-b1KhuryA
2 *A voice among the ashes*.
3 Serpente Infinita (*Infinite Serpent*), Respirar de Ouvido, 2018. Listen here: https://open.spotify.com/album/7w5kA5fqBhzqrVYHRip5rT?si=-LWmud41RgiRPJl NindoWw
4 Constel is an instrument invented by the Brazilian percussionist Leandro César. It consists of a table with several piano strings of different lengths and thickness.
5 This idea of exploring the note through varying the speed and direction of the bow is somewhat related to the music and playing of Andrew Dreyblatt.

6 Eventually the dense, irritating sound would stop to give place to the lyrics of 2Pac's Ghetto Gospel: *"Before we find world peace/ We gotta find peace and end the war in the streets/ My ghetto gospel"*. 2pac. Loyal to the game. Ghetto Gospel. Interscope Records. 2004.
7 "Cidades Invisíveis" *(Invisible Cities)* (2012) for viola, electronics, percussion, and field recordings was premiered and performed as part of the conference "Ler, ver e ouvir Cidades Invisíveis" with PhD Architect Nuno Grande (FABAUP) and PhD Rita Marnoto (FLUC). TV coverage can be found here: https://www.youtube.com/watch?v=JtiNa2fTwjw.
8 To find out more about the creative process of "Invasão", read my Doctoral thesis "Pensamento Musical: a composição como processo", digital repository of the University of Coimbra, 2016.
9 "Invasão" (*Invasion*) was premiered at the Djerassi Residency Artists Program, California, USA, in June 2012. Listen to excerpts of the premiere here: https://josevalente.bandcamp.com/album/invas-o.
10 "Alto do Minho" is a documentary film by Miguel Filgueiras: https://vimeo.com/32779931.
11 "Serpente Infinita" *(Infinite Serpent)* can be listened in any digital platform: https://open.spotify.com/album/7w5kA5fqBhzqrVYHRip5rT?si=4sZJC2UnSuiXZs8SMR0LUA.
12 "Circulando" (*Circulating*) is an art installation created together with visual artist Carlos Mensil. It was premiered at the exhibit "Síntese Activa" at Fórum Arte Braga in July 2019. https://carlosmensil.com/circulando-2019/.
13 *Interrupted Respiration.*
14 Chewing and Every day.
15 Relationships between individuals.
16 Tisana means tisane or herbal tea.
17 "Once upon a time there was an infinite serpent".
18 "Once up on a time there was an infinite serpent, / because it was infinite there wasn't a way to / know where its head was. / Each time you took a vertebra (n.f.t: from the snake) / it didn't need any. You could even / break it, dislocate it or mend it. It would remain / forever infinite. Whoever would want to / take a piece home (n.f.t: from the snake) / could put it on the wall and contemplate / a fragment of the infinite serpent." Hatherly 2006: 50.
19 Hatherly 2006: 46.
20 Hatherly 2006: 54.
21 Hatherly 2006: 51.
22 Truce.

References

Barenboim, Daniel. *Está tudo ligado: O Poder da Música* (Lisboa: Editorial Bizâncio, 2009).

Bergson, Henri. *O Riso* (Relógio d'Água Editores, 2009).

Cabanas, Edgar. Illouz, E. *A ditadura da felicidade* (Temas e Debates. Círculo de Leitores, 2019).

Cage, John. *Silence: Lectures and Writings* (London: Reprinted by Marion Boyars Publishers, 2009).

Calvino, Italo. *As Cidades Invisíveis*, 12th ed. (Lisboa: Editorial Teorema, 2010).

Camus, Albert. *O Mito de Sísifo* (Lisboa: Livros do Brasil. Porto Editora, 2016).

Domingues, Álvaro. *Vida no Campo* (Porto: Dafne Editoras, 2011).

Dorfles, Gillo. *Oscilações do gosto: a arte de hoje entre a tecnocracia e o consumismo* (Lisboa: Livros Horizonte, 1974).
Figueiredo, César. *Frank Zappa: a Grande Mãe* (Porto: Fora do Texto. Centelha, 1988).
Han, Byung-Chul. *A Sociedade do Cansaço* (Relógio d'Água Editores, 2014).
Hatherly, Ana. *463 Tisanas* (Quimera, 2006).
Levi, Primo. *A Trégua* 1st ed. (Alfragide: Publicações Dom Quixote, 2017).
Oliveros, Pauline. *Deep Listening: A composer's sound practice* (Deep Listening Publications, 2005).
Peirano, Marta. *O inimigo conhece o sistema* (Faktoria K. Kalandra, 2020).
Pereira, Nuno Teotónio. *Escritos (1947–1996 selecção)*. 1st ed. (Porto: Faculdade de Arquitectura da Universidade do Porto, 1996).
Pereira, Ricardo Araújo. *A Doença, o Sofrimento e a Morte entram num bar*. 8th ed. (Tinta-da-China, 2019).
Rancière, Jacques. 2010. *O espectador emancipado*. 1st ed. (Lisboa: Orfeu Negro, 2010).
Rieman, Rob. *Nobreza de Espírito: Um Ideal Esquecido* (Lisboa: Editorial Bizâncio, 2011).
Rieman, Rob. *O Eterno Retorno do Facismo: de eeuwige terugkeer van het facisme* (Lisboa: Bizâncio, 2012).
Seabra, Augusto. *António Pinho Vargas: Grafitti (Just Forms); Six Portraits of Pain; Acting Out* (CD booklet) (Porto: Casa da Música, n.d.).
Valente, José. *Pensamento Musical: a composição como processo* (Coimbra: Tese de Doutoramento, Universidade de Coimbra, 2016).

Filmography

Chomsky, Noam; Foucault, Michel. *Debate Noam Chomsky & Michel Foucault – On Human Nature*. 1971. (Available on Youtube. 2013.)
Filgueiras, Miguel. *Alto do Minho*. 2012.
Grande, Nuno. *Sagrado*. Porto: Ruptura Silenciosa. Faculdade de Arquitectura da Universidade do Porto. 2014.
McFerrin, Bobby. *Bobby McFerrin from USA – Interview*. Beatbox Battle TV. 2010.
Zappa, Frank. *Frank Zappa interview for TROS TV Show 1991*. Not available. 2012.

Discography

2pac. *Loyal to the game*. Interscope Records. 2004.
Ambarchi, Oren. 2016. *Hubris*. Editions Mego.
Arf Arf. 2013. *Arf Arf*. Tochnit Aleph.
Borodin Quartet. 2018. *Shostakovich: Complete String Quartets*. Decca Music Group Limited.
Camus, Albert. O Mito de Sísifo (Lisboa: Livros do Brasil, 2016).
Carvalhais, Hugo. 2015. *Grand Valis*. Clean Feed Records.
Children of Bodom. 2015. *I Worship Chaos*. Nuclear Blast GmbH.
Contreras, Billyand and Howes, Cristian. 2004. *Jazz Fiddle Revolution*. BRC: a division of Cojazz Recordings, Ltd.

Douglas, Dave. 2016. *Dark Territory*. Greenleaf Music.
Emperor. 2003. *Scattered Ashes: A Decade of Emperial Wrath*. Tanglade Ltd t/a Candlelight Records.
Ensemble Intercontemporain. 1998. *Berio: Sequenzas*. Deutsche Grammaphon GmbH, Berlin.
Guy, Barry and Evans, Peter. 2018. *Syllogistic Moments*. Maya Recordings.
Knox, Garth. 2009. *Viola Spaces*. Mode Records.
Kopatchinskaja, Patricia. 2019. *Time & Eternity*. Alpha Classics/Outher Music France.
Korn. 2016. *The Serenity of Suffering (Deluxe)*. Roadrunner Records.
Léandre, Jeanne. 2005. *Concerto Grosso: live at Gasthof Heidelberg Loppem*. Kunsthallelophem.
Martin, Luis José. 2017. *Tentos – invenções e encantamentos*. Shhpuma.
McFerrin, Bobby. 2010. *VOCAbuLarieS*. Universal International Music B.V.
Opeth. 2016. *Sorceress*. Moderbolaget Records under exclusive license to Nuclear Blast.
Ostertag, Bob. 2013. *Bob Ostertag plays the Aalto*. Self-edited.
Saldanha, Jonathan. U. 2016. *Tunnel Vision*. Silo Rumor.
Slayer. 2015. *Repentless*. 2015 Nuclear Blast GmbH.
Shyu, Jen. (n.d.). *Song of Silver Geese*. Pi Recordings.
Tamestit, Antoine and Widmann, Jorg. 2018. *Widmann: Viola Concerto*. Harmonia Mundi.
Valente, José. 2015. *Os Pássaros estão estragados*. Coimbra: Jacc Records.
Valente, José. 2018. *Serpente Infinita*. Serpa: Respirar de Ouvido.
Various artists. 1997. *Constança Capdeville*. Miso Records.
Zappa, Frank. 1993. *Roxy & Elsewhere*. Rykodisc.
Zappa, Frank. 1971. *200 Motels*. Rykodisc.
Zíngaro, Carlos. 2008. Regef, D., DeJoode, W. *Spectrum*. Clean Feed Records.

PART II
Collaborative writing experiences (from the field)

8
CREATING ATMOSPHERES WITH LL

The forms of creativity of Tiago Oliveira

Melina Aparecida dos Santos Silva, José Cláudio Siqueira Castanheira and Tiago Oliveira

Introduction

On the first day of fieldwork in Luanda, on August 29, 2017, one of the authors of this chapter had the opportunity to go to a rehearsal of *Kosmik*, the Angolan progressive rock band formed by guitarist Ricardo Carmo, drummer LP, and keyboardist/bassist Tiago Oliveira, who is also a music producer at Studio 2 (Silva 2022). The rehearsal was held at the Brazil-Angola Cultural Centre, the venue for the group's performance that would take place on September 1, 2017. While watching the rehearsal, which followed a dynamic close to that of silent cinema, i.e., the trio performed the instrumental compositions, which served as a musical accompaniment for projected images, she wondered how Tiago could handle so many functions during a performance.

We can relate Tiago's creative versatility, whether in *Kosmik* or Studio 2 (located in downtown Luanda), to his first acquisition of a loop pedal in 2010. However, Tiago was already acquainted with digital looping procedures since he began to make music using synchronisation between a computer and other musical instruments:

> One way or another, my initial form of composition, since I had no musical basis, ended up very much around creating loops with a computer, but I worked a lot with rhythmic and melodic sections that I created, putting them in loops and building themes from there.
>
> *(Tiago Oliveira, February 28, 2021)*

Although this first model had fewer functionalities than his current live loop pedal, Tiago considered it more "intuitive". Therefore, the pedal has

DOI: 10.4324/9781003154082-11

accompanied him longest as a tool for musical composition and for the most personal musical performances. Among the already existent musical projects in which Tiago has taken part, we can mention *Dinamite Napoleon*, a solo experience more focused on electronic music and rock, and *Estranha Trator*, a power trio of industrial rock and heavy metal with instrumentalists Zé Beato and Miguel Gulander, from the city of Lobito. In 2013, Tiago was invited to participate in the band *Hiflow*, dedicated to funk and soul music, with a proposal, which was, in his words: "more commercial than I am used to listening to".

After the opening of Studio 2, in March 2016, and the most intense contact with local bands, Tiago set up the *Kosmik* music project with Ricardo Carmo and LP. For the instrumentalist and music producer, the aim of the Angolan group is the focus on progressive and "psychedelic" rock compositions. However, when analysing his artistic trajectory, Tiago did not consider himself a musician in the strict sense of the term:

> If you compare me to musicians, I'm usually the least musician-like of all. I even consider myself more a painter than a musician. I have a special attraction to build landscapes and moods through sound. I am not a technician, or an expert on any instrument. I make music with instruments, but I have never perfected the technique in any of them. However, I have some ease in creating atmospheres. This turns out to be a common element in the projects in which I participate. It is in this area that I end up having a greater contribution: in the landscape part.
> *(Tiago Oliveira, September 20, 2017)*

Tiago's argument is quite significant because, besides framing the use of looping devices as a somewhat distanced practice from traditional musical performance, it also leads to the question of his own development as a musician or even the role of musicians in live performance.

Thus, the purpose of this chapter is to discuss the function of live looping (LL) as a form of creating "sound atmospheres", an aesthetic resource commonly used by Tiago in his musical projects. Qualitative methodology was used in the research, with data collected from an interview with the instrumentalist and producer carried out via email on February 28, 2021, and based on a bibliographic review about LL. In the first section of this article, we will address discussions about looping practices and possibilities, and their impact on the composer's performance. In the second section, we will present two performative analyses of Tiago's aesthetic proposals, describing his dynamics with instruments such as synthesisers and LL pedal to build layers and ambiences in the repetition of sound excerpts.

Looping and performance

In an interview granted to the authors, Tiago explains his interest in looping from the acquisition of his first LL pedal. He draws attention to the fact that the tool helped him in his solo work, and was not related to a specific group. The notion of organising recorded excerpts in loops, later adding other musical elements, was already part of his composition practice. In this process, Tiago used the computer as the main tool. The use of loop pedals brought different dynamics to his work:

> Later, perhaps around 2010, I'm not sure, I was already in Angola and had the opportunity to buy my first looper and, from then on, it became a tool that I used for some small musical performances, but it remains a working tool to compose music.
>
> *(Tiago Oliveira, February 28, 2021)*

The use of pedals and other small electronic gadgets offers great possibilities for live performance, but also imposes some restrictions regarding real-time decisions in the interaction with devices. In general, musicians play a musical motif (rhythmic or melodic) live for the first time, recording and then replicating it continuously, creating a base for the subsequent elements to be added later. This procedure implies at least two specific issues. Firstly, the use of a pre-recorded base with factory audio samples, a very common practice since the development of software-based compositional and recording tools, is no longer such a determining issue. There is a change of focus in terms of who "makes" the sound; the performer must produce the samples himself, which re-signifies his own presence on stage. This topic will be returned to later in the chapter. Secondly, unlike the pre-programmed bases using ready-made samples, the creation and live execution of loops requires specific training and virtuosity – different from the training of traditional musicians – and that recreates the relationship between the musician and the technological apparatus he depends on.

Analysing electronic device-mediated gestural performance, Sarah Nicolls (2010) highlights the importance of improvisation to deal with the difficulties that traditional musical notation (and classical theory in general) has with describing the process of live creation:

> The individuality of each of these performers only strengthens the case that improvisation is not only a way of generating music but also the key to inventing and learning a host of new instruments, interfaces or systems of interaction.
>
> *(Nicolls 2010: 4)*

The musicians' gestures are reshaped to the extent that their interaction does not only take place with the instrument they play, whether or not it is adapted to some type of responsive apparatus (such as EMG sensors, which read the electrical current created by muscle contraction), but also with the set of devices that, in addition to the instrument itself, also produces real-time changes in the generated sound. These effects, at least most of them, are controlled by the performer's gesture. The gesture thus instructs the instrument, triggering notes, but it also instructs connections that read movements, pulsation, and electrodermal activities that, in turn, produce or change the sounds initially predicted. Being, to a certain extent, unpredictable, such bodily responses demand a new relationship between the sound that is intended and the one that is obtained. In other words, there is greater scope for discovery, experimentation and improvisation. Another aspect of great importance is that the gestures, thus enhanced, become the target of audience interest. The experience of listening is inseparable from the experience of observing the performers and their physical movements. The artists' body as a whole is made the place of musical creation, and not just their intellect or the partial training of hands or lungs for playing instruments.

> Intelligence and creativity is actually written into the artist's muscle and bones and blood and skin and hair.
> *(Ostertag quoted in Nicolls 2010: 48)*

Chapman (2013) reminds us of how physicality was limited, within the bourgeois perspective of European culture, to the very least necessary to accomplish the precise note. As an object of worship, musical sensitivity – of both listeners and instrumentalists – should transcend the objects with which music was created. Musical writing was, likewise, the abstract representation of a physical phenomenon and, thus, represented the evanescent air of artistic intelligence. Street musicians, who often attracted attention by playing various instruments at the same time (guitar, harmonica, percussion, etc.), brought to the fore the not so noble side of musical experience, leaving such symbolism aside. Their performance was mesmerising because of their almost acrobatic skills, and because of the use of unusual paraphernalia that would allow them to control more than one instrument at the same time. The creativity to solve physical limitations during playing presents, as a result, an expanded performance (or a joint performance of musician and machines). More than the instruments, the machines designed to allow musicians to play different instruments with various parts of their body became the centre of attention. The musician's virtuosity (as a technique perfected by traditional studies) is obscured by the dexterity of combining and making unorthodox arrangements of various objects work perfectly. This type of hybrid performances, where the artist's creativity is not seen just as something autonomous and immaterial, brings out the material aspects of music, especially the relationship between the musician's body, instruments and other unusual tools. Therefore, they were considered by the elitist artistic circles as manifestations of lesser importance, not really artistic:

> At the same time, this situation could not be more distant from the conception of music idealised by the bourgeois critical establishment of that time, which had come to celebrate music as an autonomous art form, its aesthetics kept meticulously separate from the materiality of everyday life.
> *(Chapman 2013: 454)*

Chapman (2013), when analysing the work of composer, instrumentalist and singer Teresa Andersson (songs *Na Na Na* and *Birds Fly Away*),[1] names her proposal of connecting music creation to live interaction with the different devices which help to create her sounds as "distributed subjectivity". Chapman says, "In live, solo multi-instrumentalism, the visible performance of multivalent skill is as important to the overall effect as the 'finished product' of musical sound" (2013: 467). Recorded music, especially from the historical formulation of studio protocols, shows a sort of opacity that is often seen as one of its virtues. Not only did recordings abandon live collective recording techniques due to greater practicality and a better technical finish – as well as a performance with fewer errors – but also because of the adoption of different resources for the control and modification of sounds. The practices of "construction" of sounds maintain a narrow relationship with technical objects and with those professionals who, even though they are not trained as musicians, demonstrate enough knowledge of how to operate such machines, as seen in Tiago's creative trajectory. The producer (or sound editor, or recording engineer, etc.) has turned into a fundamental component in the creation and recording of songs as the very definition of music became more fluid through the years.

Thus, we can analyse performances using this specific type of technological elements as a resizing not only of the human-machine relationship, but of the compositional architecture itself that, as Marchini et al. (2017) remind us, adapts to a type of functionality (and limitation) present in the devices used (in this case the loopers). The complexity of the movements required for the real-time production of music, no longer from previously recorded samples, but from live performance fragments, made it necessary to adapt the harmonic structure so that it can function cyclically, and for it to accommodate the successive layers added by the musicians, requiring just the minimum necessary activation of equipment controls.

Maintaining the tonal orientation common to the greater part of Western music, diverse looping-based experiments adopt closed harmonic cadences as a central strategy. Despite an affinity with other types of composition in terms of instrumentation and melodic lines, these projects differ somewhat from the basic structure of most pop songs, i.e., they avoid alternating main parts and chorus, demonstrating a certain degree of variability. Thus, much of the solo musician's performance accompanied by looper and sequencer parts can be characterised as an exercise in completing a predefined and relatively limited base. Tone transpositions or time changes (relatively simple actions for groups of musicians playing live together) are complicated procedures and to some extent not within the scope of that kind of work. Repetition (always slightly different because of the addition of new layers) becomes the core of the composition and has an additional attractive element: the musician's ability to organise – in real time – all musical elements and technical steps. In the case of *Na Na Na*, Teresa makes a break in the main motif, inserting a bridge – with a significant reduction in the number of sound layers. After that temporary shift, she returns to the main part of the song. That kind of variation requires of the artist some complex training regarding gestures in general and the control of her setup. One possible solution, in cases such as these, is to expand the setup to meet the most complex dynamics of the song; that is the solution in Teresa's case, but in many other examples live performance is restricted to creating on a constant harmonic base repeated by a single looper.

When we think, therefore, about the notion of improvisation, we abandon the field that deals exclusively with the making of musical notes to incorporate other types of sound expression. The performer improvises with the superimposing of loop layers and with the production of specific live timbres. Effects pedals, despite offering many configuration options, are still subject to the unpredictability of live performance. In this sense, the relationship with the apparatus enters the realm of improvisation, both by accident and by conscious experimentation. The risk of error, whether in coordinating different equipment, or due to any of them not functioning correctly, leads the artist to a terrain of uncertainties, where they must anticipate each of their actions and also the combination of the resulting sounds. Unlike live performance with a band, the negotiation between the various musical elements is not so simple, being built from the interaction with "deaf" devices, endowed with invariable and unfailing playing ability, but unable to "read" or "anticipate" the artist's intentions. In this sense, with the purpose of deepening the notion of "distributed subjectivity", mentioned by Chapman, we propose the existence of a principal human subjectivity that needs to control the subjectivity of the machine. The human element suffers multiple constraints from the various automated objects during the presentation, and if it does not demonstrate the mastery over the functioning and logic of the apparatus, the performance is not a success. The artist's subjectivity must respond by following patterns produced by the machine, but it should be kept under control. In Chapman's words: "the entrepreneurial self of the one-man band navigates a world in which the conspicuous performance of flexible dexterity counts for as much as its end result." (Chapman 2013: 461)

Traditional spaces for music production, such as music studios, have lost part of their centrality as places of sound creation with the popularisation of loop-based performance. Historically, the studio began to represent, for musicians of different styles, the opportunity to build a technically suitable "sound". Thinking about performance with loopers as a gestural live construction of sounds, the idea of technical adequacy, at least according to the model of studio recording, has some of its strength diminished. On the contrary, what emerges as a driving principle in this type of performance is the ability to organise and anticipate musical movements from the perspective of repetition machines. The body of the musician needs to connect itself with instruments that, although limited in their programming, have their own sense of autonomy when in operation.

We now move on to the analysis of Tiago's performance on two of his online videos.

3 AM Sessions and Sci-Fi/Ambient Improvisation

Tiago's improvisation with looping techniques was uploaded to YouTube in November 2013. The performance has 5,371 views, 48 likes, three dislikes and 21 comments. The song *3 AM Sessions* is 09:49 long and was composed using a laptop with Ableton Live software in "Session" mode, a Nord Lead 4 synthesiser with two voice oscillators, a four-string Fender Precision bass, and a Korg Volca Beats Drum Machine. The drums used during improvisation were pre-programmed in 4/4 measure using kick and snare samples. At 00:08 of the performance, Tiago records the bass line, creating a loop of four compasses, which constitute the basis of the composition.

Subsequently, at 1:22 of the video, Tiago creates delicate sound textures on the synthesiser, which are used as a form of ambience for the following improvisations. He continues in the high pitch notes of the bass scale to create new melodic lines and adds to these same tracks the ambience effect created on the synthesiser. Tiago performs a bass solo from 03:54 until 4:59, which is then put into looping as a new sound layer. According to Latartara (2010), composing with a laptop as the primary instrument is associated with cultural codes of electronic music, one of Tiago's most valued musical genres. Although the use of a laptop for composition does not imply genre and/or musical style norms, since the 1990s, the possibilities created by such technology, like portability, have fascinated new composers. They superimpose sound layers with repetition and noise elements that will be used throughout the musical cadence and add small looping parts (Latartara 2010), as seen in Tiago's performance. The title *3 AM Sessions* is suggestive of the creation of ambiences in the musical piece.

Repetition and noise elements help us understand the relationship between the compositional structure, the automated musical instruments and the aesthetic proposals of each composer, as evidenced by another Tiago performance entitled *Sci-Fi/Ambient Improvisation* uploaded to the YouTube channel *From Stereo* on July 20, 2020. Although this recorded performance is simpler compared with the construction of the melodic lines in *3 AM Sessions*, the composition invites us to have a better appreciation of Tiago's sound proposals. *Sci-Fi/Ambient Improvisation* was created using a Roland JP-08 with the portable keyboard K-25m, Volca Beats, P-Bass and Ableton Live to sequence. The JP-08 Roland emulates the sound and controls of the iconic synthesiser Jupiter-08, a discontinued model of the same company. At 07:21, Tiago uses the synthesiser as a way of creating ambience, e.g., as a sound representation of the wind, not concerning himself with melodies. The accumulation of sound layers actually gives us the impression that we are inside a science fiction narrative. The dialogue of bass and synthesiser acts as transportation between parallel worlds, as a picture of unknown spaces and as a close contact with the supernatural. Tiago's improvisation could easily be taken to be a musical score from sci-fi or terror audiovisual productions of the 1980s, as his creativity intensely explores the mysterious and futuristic feel of synthesiser tones.

The production of improvisation sound textures is similar to that of *3 AM Sessions* with pre-programmed electronic drums and repeating looping parts. At 01:29, still working in the 4/4 metric, Tiago inserts electronic tones playing the role of a closed hi-hat. At 01:54 he finally plays the bass part. With the bass line and electronic drums already playing in a loop, he starts modifying some sound wave parameters and adds a more complex beat to the composition.

What drives our attention is the practical effect of using the pedal in the performance as a whole. This device, dedicated to the repetition of the small recorded live sections, not only being very functional in terms of the elaboration of loops but, at the same time, having itself great control over the musician's performance. Latartara (2010), in his analysis of composers Markus Pop, Masami Akita, and Miguel de Pedro, reflects on the direct influence of the human-computing interaction in the compositions: "Layers of repetition or loops can be viewed as reflecting a Virilio-like machine aesthetic displayed in the musical structure through the software interface used to create the music" (Latartara 2010: 113).

In this sense, the freedom that the musician acquires in this new relationship not only favours improvisation but requires a capacity to improvise that differs from that of jazz. Marchini et al. (2017) point out that the potential of Reflexive Looper[2] has not yet been properly explored due to the general focus that research has on jazz improvisation. "Unlike a common loop pedal, each layer of sound in RL is produced by an intelligent looping-agent which adapts to the musician and respects given constraints, using constrained optimization" (Marchini et al. 2017: 139).

In a mapping of 145,000 LL videos on YouTube, Marchini et al. (2017) noted that the performers used the repetition of a fixed sequence of four chords for the amateur artist to sing or perform guitar solos with. However, there were few cases where composition structures had breakdown sections to start new chord sequences. According to Marchini et al: "This occurs because the common loop pedal makes it difficult to produce transitions between sections with different durations and/or chords. As a consequence, only songs with static harmony are performed with a pedal loop" (Marchini et al. 2017: 139). In Tiago's case, he is constantly concerned with the construction of melodic lines and gloomy sound textures from bass-built repetition loops.

Conclusions

This chapter addressed the forms of creativity and the musical structures developed by the producer and multi-instrumentalist Tiago Oliveira in his works *3 AM Sessions* and *Sci-Fi/Ambient Improvisation*. Both pieces explore the features of looping parts combined with the creation of soft textures on the synthesiser and melodic lines on the bass to achieve sonic ambience for Tiago's improvisations. Although looping techniques have specific features and configurations for building repetitions of motifs, which can give the impression of similarity between sound layers in his pieces, Tiago's creativity is directed towards the possibility of creating new forms of musical atmospheres. More than using loop repetition features to organise sound elements and live performance procedures, Tiago uses looping as a means of composing songs to be specifically performed in his musical projects. Repetition and noise elements help us understand Tiago's creative premise about musical structure and instrumentation.

The use of repetitions provided by this type of device has a clear influence on the delineation of an "aesthetic" of loopers, giving compositions a vertical dimension, with the continuous overlap of new sound elements. The horizontality of the songs, from this perspective, is limited to a virtually unlimited flow of bars that, by the nature of repetition, do not move to a conclusion. With a certain minimalist approach, the performance lasts until the musician feels that all the necessary sound elements have already been presented and, as a result, the song is complete. Thus, harmonic immobility can be seen from the point of view of a contemplation through the musician's gestures, where the machine's task would be to support the themes presented at each cycle, functioning as a recurring memory on which the performer – improvising in many cases – develops his statements. More than accumulated notes, performance with the aid of loopers is a reflection with and about musical gestures.

Notes

1 *Na Na Na*, available at: https://www.youtube.com/watch?v=n2eD4GcLohE.
  Birds Fly Away, available at: https://www.youtube.com/watch?v=vMXqn42AykM.
2 Mancini et al. call *Reflexive Looper* (RL) the live-looping system where a solo musician is able to build a whole rhythm section, chord progressions, bassline etc by looping live-played excerpts of music.

References

Chapman, Dale. "The 'one-man band' and entrepreneurial selfhood in neoliberal culture. *Popular Music* 32, no. 3 (2013: 451–470. https://www.jstor.org/stable/24736784.

Fenn, John. "The building of Boutique Effects Pedals – The 'Where' of Improvisation". *Leonardo Music Journal* 20 (2010): 67–72. https://www.jstor.org/stable/40926375.

Latartara, John. "Laptop Composition at the Turn of the Millennium: Repetition and Noise in the Music of Oval, Merzbow, and Kid606". *Twentieth-Century Music* 7, no. 1 (May 2010): 91–115. doi:10.1017/S1478572211000065.

Marchini, Marco; Pachet, François, & Carré, Benoît. "Rethinking Reflexive Looper for structured pop music". Proceedings of the International Conference on New Interfaces for Musical Expression, 2017, 139–144.

Nicolls, Sarah. "Seeking Out Spaces Between: Using Improvisation in Collaborative Composition with Interactive Technology". *Leonardo Music Journal* 20 (2010): 47–55. https://www.jstor.org/stable/40926373.

Silva, Melina Aparecida dos Santos Silva. *We do Rock Too: Formas de criatividade do rock angolano* (Rio de Janeiro: EDUERJ, 2022).

Links

Oliveira, Tiago. Looping/Improvisation – "3 AM Sessions" – Nord Lead 4 + Korg Volca Beats. YouTube, Nov. 7 2013. https://www.youtube.com/watch?v=T1CQWhVlsI0&list=PLnfJynLS0oFkET8ZxiN1o_VJG-JbDg8sJ&index=1.

Oliveira, Tiago. Sci-fi/Ambient Improvisation (Roland JP-08, Korg Volca Beats). YouTube, Jul. 20 2020. https://www.youtube.com/watch?v=A8e7ubBzOno&list=PLnfJynLS0oFkET8ZxiN1o_VJG-JbDg8sJ&index=7.

9
THE LOOP PEDAL AND THE GUITAR – INDIVIDUAL PRACTICE AND ITS CONNECTION TO SOCIALITY

Ricardo Jorge Monteiro Cabral and Jorge Vicente dos Santos Almeida

The young Cape Verdean guitarist

When you see Jorge performing, you feel like you are navigating through time. A musical progression guides the listener in experiencing a musical beginning, middle and end. Time becomes inconsistent and untraceable through a clock, but it perfectly fits the listener's perception linked to the subjective musical experience. Rhythms, beats or dynamic transitions are the only way to keep track of time. Jorge often closes his eyes, not because he fears being judged by others' gazes, but because he connects his musical performance to personal emotions. He knows what he wants, but he is open to being surprised by his mastered execution and equipment possibilities. His head shakes back and forth for a few moments, conveying a sense of confidence. He points his face upwards at other times, almost as if that confidence is in danger of being swept away by some musical detail of perfection. Still, he embodies the readiness and precision of a seasoned performer with a hint of unpredictability and emotivity. The rational and irrational coexist in the same space, mediated by Jorge's body movements, a guitar, and a loop pedal.

"I believe that I can achieve a unique and profound experience from the quality of my equipment", he asserts. The interplay between the pedals and his guitar produces a musicality that makes him achieve meditation. The movements of his feet, arms, and hands create the illusion of a dance that seeks solid individual freedom. He emphasises this concept by saying, "wherever I turn on my equipment, I feel that that is the only place I want to be". The perfect combination of art, movement, props, technique and knowledge makes him feel safe. He sees himself as able to ponder on personal things when he records and performs his loops. He repeatedly refers to his essence, pursuing himself through his gear and the music produced. Being spiritual, he believes that when his equipment and sounds surround him, he is closer to God – whom he considers his best friend and who inspires him.

Jorge Almeida was born in São Vicente, an island in the archipelago of Cape Verde, in October 1990. Among family and friends, he goes by Djodje. He grew up in Rua do Douro, commonly known as Monte. Like most local youths interested in music, his first musical explorations were purely physical experiences in which the body was the primary instrument. He remembers making beats by patting his legs with slippers in order to follow the rhythms that he could hear from songs broadcast on the radio or played on television. The eagerness to challenge himself by creating rhythmic patterns developed Jorge's musical techniques from an early age. Another experience with a significant impact on him was when playing the flute. The idea of producing a musical note was exciting to him. The constant musical emulation came up again, and replicating songs known to him became enjoyable. Jorge's talent did not go unnoticed and his mother, Maria dos Santos, bought him his first guitar so that he could explore further. Being a key instrument in Cape Verdean music, guitars introduced young Jorge to music genres such as Morna, Talaia Baixo, Funaná, and Coladeira. Living in Boavista, another Cape Verdean island, led him to learn and improve his guitar skills with his friend Robbie Brito – a local musician. He would create an unbreakable bond with the instrument from this point onwards.

The loop pedal came into his life later. By desiring to improve his guitar sounds and performance, Jorge aimed to find a delay effect processor – to be added to the guitar output and song. The first time he saw a loop pedal was when he went to buy a delay effect pedal. Jorge purchased a loop pedal that could produce delays in his performance. He didn't know what a loop was, but he soon realised what he had come across. He then started to use the loop function on the new machine. From that moment on, the young guitarist started using it resolutely, recording and reproducing his performance; this allowed him to add on to a previously recorded cluster of sounds, or loops, having a constant overlapping of audio layers in real-time. The practicality of the loop brought Jorge endless possibilities.

It took work for Jorge to acquire the pedal. A good deal of electronic equipment is not standard in Cape Verde, nor is it in Cape Verdean culture. There are no national artists who have guided Jorge to master the loops. Until recently, no sources, such as the internet, taught the art of looping through videos shared by foreign artists. No musicians used pedals, mainly because they were hard to find. It is important to note that Cape Verde is an example of a rural country with very few shops interested in sophisticated equipment. Cape Verde still has traditionalist views surrounding music. Although you can find a variety of genres such as Rap, Rock and Jazz nowadays, the conservative idea of performing and creating traditional music is resistant.

Playing Morna using a modern electronic extension, such as a loop pedal, can be very misunderstood, and Jorge has examples of such misunderstandings in his journey. One example was when he started playing on the island of Boavista, in tourist hotels, where, while he played music, a young Cape Verdean man came to him and accused him of cheating. The young man truly believed that Jorge was pretending to play and that a mysterious machine was covertly reproducing tracks with the guitar's sound. Jorge's embarrassment that day made him develop a sense of self-consciousness when presenting himself as a music artist. He began to show the details of his playing system and unusual material to those in charge of the events in which he participated. "I explained what I do musically, so they wouldn't think I was using a pre-recorded track. I would show them everything I was doing was in that moment", says Jorge. The young artist did not give in, valuing the creativity provided by the loop pedal. The pedal became an instrument with traditional possibilities. For him, the small piece of equipment he steps has become the redefinition of what it is like to live in the Cape Verdean music environment and be, to a certain extent, Cape Verdean.

Equipment and music logistics

Jorge's innovative sound developed from a new music system. He has been inspired by international figures such as Ian Ethan Case[1] – an artist who combines stringed instruments and loop pedals (Ianethan.com n.d). Jorge's sounds require the recognition of the loop station technology to optimise the music in a specific duality: connecting with traditional aesthetics and exploring new artistic characteristics. Jorge does not usually employ a mixing desk when composing alone. His guitar, which has an internal microphone and a jack input, sends the sound to a simple pedal with the ability to equalise and add reverberation and echo. This can also be done through an amplifier. After this initial process, the guitar's sound reaches the loop pedal, where, in addition to effects similar to the first device, it also contains the loop effect. With this looping device, the artist can record performance sections, creating chunks of sound and overlaying them, making it possible to initiate an elaborate form of musical production (see Duarte 2020: 10–16). The last step is to send the sound to the speakers – or a mixer in a stage performance. Jorge shapes the guitar's sound, fragmenting it and making textures (see Schloss 2014: 136) while reproducing it, which has become his trademark. His execution and creation require knowledge and logic: knowing where the sound is coming from, where it goes, and what will happen at each stage. This logic guarantees Jorge a way to predict his outcomes, and he only makes mistakes when he forgets the sense of the natural flow between sound and the sequence of actions, processors, and cables. Moreover, technical knowledge is developed to predict how the sound will be produced through the specifics of guitar playing. The mixture of different processes, performances and effects must be combined with the aesthetic imagined by the artist and how it connects to his cultural and social background.

The execution with the loop pedal is undoubtedly a process involving exceptional coordination of the legs and feet (Duarte 2020: 16). The act of stepping on equipment to activate or deactivate specific audio effects, or the insertion of another musical take, requires readiness, assertiveness, and muscle memory. The performer's mind must be trained to recognise the musical timings and synchronise them with the timings of their muscular reaction and the time necessary for the pedals' audio processing. This synchronisation happens only after having mastered a seamless use of the station, together with good interaction between mind, body, music and the gear itself. Thus, his music process is generated by the precise mechanical timings and the musical tempos (see Duarte 2020: 16–17).

Jorge explains this by saying that he considers his feet functioning as if he was a drummer stepping on the bass drum. The constant tempo at which the bass drum is usually played, having exact numbers of attacks fulfilling the musical sequences and distributing attacks at equal intervals, is an excellent example of understanding this process. It means that, by having some percussion logic, Jorge can understand and identify sequences and characteristics to which musical sounds respond. This principle involves the possible timings for the beginnings or ends of sounds coming from any musical instrument. Inspired by the likely beats of the bass drum in a sequence, artists like him know where loops should begin and end. That means that Jorge also knows when to step on the pedal or not.

The possibilities of using a loop pedal extend beyond the realm of music creation into rehearsal activities. Using the pedal to create loops makes it possible for the rehearser to build exercises on top of one looped audio base track. The base track, or first layer, should guide chord scales, making those practising exploring other melodic compositions aesthetically compatible with the previous musical scale; this gives rigour to the practitioner's actions (Duarte 2020: 11). Using a loop also makes it possible to develop the notion of possible musical lengths: the time of loop extension and its full time of repetition (Schloss 2014: 138; Girdzijauskienė 2015: 109). The loop extension will guide the artist's intention, adapting their music possibilities according to the time available from the start to the end of the loop itself. After that, the loop repetition will demand a broader perception of time: the music approach as a whole, making the artist construct arrangements that will be carried on for an extended period. Every music arrangement must be conceived and linked to the loop itself and its repetitions, at precise moments and alongside the music progression. Jorge demonstrates that when creating a loop, the more straightforward thing is to enter the first chord or note and the most challenging thing is to finish the sequence correctly. According to the maths behind the musical timings failing to complete a loop will culminate in inconsistent results, producing wrong sequences and damaging the whole musical structure. Timing is crucial on both micro and macro scales of music.

How to make sound and music

Both guitar and loop pedal accomplishments require training time. Jorge admits that he has the luxury of being able to practise almost every day. He clarifies that musical development is impossible at the performing and creative level without investing in daily exercise. Training and repetitions are essential because they help memorise physical implementations, exercise the muscle to perform in specific ways, and allow musical creativity to flourish. Creativity is achieved when the muscles and mind are trained, making them act almost automatically in sync, allowing space and energy for mental processes that lead to inspiration. Thus, ideas are materialised through body capacity when using musical instruments or equipment, and body capability should be seen as musical characteristics. In other words, to be incredibly creative, the artist has to be outstanding and coordinated in his body performance to the extent of automation.

A person who performs well with the pedal and guitars will also have more potential to explore and gain musical ideas from these objects. It is necessary to have the discipline for creativity and quality to become a reality. Jorge believes this discipline cannot be disconnected from one's living. He supports this statement by referencing his own life, stating that he would be limited both musically and in his professional relationships if he had no rehearsal routine. From this, we realise that daily discipline and the relationship with society are connected. Jorge has invested in his pedals and guitars, which allows him to please event management companies, other artists, and listeners, inducing them to rely on him and his potential. A personal view of life drives his attitude about the relationship between musical improvement and better social connections. Jorge spiritually believes in effort and gaining from this relationship between the musician and society. For him, each of his musical contributions is a gift to humanity, and the latter will pay him back with accommodation and support. Making daily loops is, in his opinion, the way to have legitimacy in the form of social status, which gives him a sense of social responsibility. A musician who owns his ability is also a musician with social skills and, thus, is an artist of universal demand.

Artists can only boost their musical ability through curiosity. First and foremost, Jorge considers himself a researcher interested in finding new solutions for his musicality. Jorge was driven by curiosity when he learned to play in São Vicente and Boa Vista's islands, purchasing his first pedal, believing he was investing in a delay machine. He speaks about an inherent interest that motivates him to produce new things and develop his abilities. Like any child who lives constantly questioning everything, his approach is a mindset of humility in the face of reality. According to him, an artist has to be curious if they want to have the possibility of yielding new musical ideas and inspirations. The same applies to working with the loop pedal; a good execution and production of loops and sequences result from the artist's openness to the machine's possibilities. The curiosity to search the options of the loop machine, the loop, and the sounds allows innovation and promising music to materialise.

The relationship between the guitar and loop pedal

For Jorge, the loop pedal is an element of mediation, deeply connecting him to his musical instruments: the guitars and, sometimes, when he individually works on the ideas for his music, the electric bass and organ. He considers his relationship with the guitar particularly comparable to someone's relationship with their best friend. For Jorge, the classical, acoustic or electric guitars – with nylon or steel strings – are much more than wood; they are his life companions. With them and a loop pedal, he creates rare sounds far from those of the natural world. In his view, the pedal is the tool that will extract the best from his string instruments. He will be able to select the best audio clips from the guitar sounds, repeating it and adding new features like the delay effect. At that point, the guitar in his hands will be free to produce other exciting songs while the sample is reproduced in a loop. But if the pedal drives sound to the best performance, shaping and adding to it, the guitar is, for Jorge, an instrument of feelings which has physical challenges and requires human effort to achieve the desired sound.

> The guitar is an instrument that is sensitive to touch. It's not a piano, in which the musical note is perfect and almost the same even if hit with a different intensity; that doesn't apply to guitars. You have to feel pain in your fingers if you want to achieve what you want. So, it's a relationship of pain and love, and you play living through it. The guitar is unique concerning contact and physical connection while seeking an emotional level. You must be dedicated and ready to sacrifice (says, Jorge).

The pain and passion needed to attain the guitar sound, combined with the mechanics and possibilities of a loop pedal, give Jorge the sound qualities he cannot renounce to this day. In Cape Verde, the guitar is an instrument valued especially in genres such as Morna. It is not uncommon to see young musicians like Jorge receive their initiation to music on this instrument. Therefore, the guitar is part of the musical identity that unites him with national traditions and aesthetics (see Turino 2010: 179). While the young artist plays his Cape Verdean-sounding guitar, when affected by the loop pedal, this culturally confined sound has the potential to be well-received across borders. With this, it is Cape Verdean music that wins new audiences. Simultaneously, the loop pedal can bring influences from outside into Cape Verdean aesthetics. The textures provided by the small loop machine are now relevant features in the cultural production of the young artist. By promoting his music, Jorge advertises the innovative musicality of a technique that has been unusual up to now. Therefore, he is an active agent in the dynamics and evolutions that are taking place in the local music and culture.

Sound, music, and inspiration

Although a pedal can, through recordings, reproduce a faithful copy of the guitar's sound, it can produce sounds that, for Jorge, seem unnatural. It happens when clipping a lousy loop, done awkwardly and off-timing, resulting in inaccuracies at the beginning or end of the captured sound. He seeks to produce a loop almost indistinguishable from the actual sound captured when recording a sample. He makes it clear by talking about his original track, *Mystic*. "*Mystic* is the best loop I have made. You don't feel the looping characteristics. It doesn't sound like a fragment. The timing of the loop click was exact", he says. The capacity to be indefinitely repeating is arguably the unique unnaturalness that Jorge accepts about loops. He is looking for the perfect imitation and precise repetition of his guitar sound when using the pedal. Due to this characteristic, it can be concluded that the loop pedal is a piece of equipment that, while distorting the reality of sounds, duplicating and overlapping them, also has the task of reproducing sounds which are exact copies of those from the sources.

One of Jorge's loop pedal characteristics is its ability to control the volumes of its audio tracks. The artist admits that he doesn't use this function much. For him, the ability to use organic dynamics, executed manually on the guitar, is more critical than the volume settings on the machine. He believes that the general music dynamics must be lively.

> Music without dynamics is only a rigid block. It doesn't have an expression. It doesn't have anything. You haven't heard anything musical if you hear music with the same structure from beginning to end. Dynamics give music a physiognomy, and it says that music has a body and curves and that it's natural and sweet (says, Jorge).

In most cases, to challenge the monotony in the broader music structure, he adds dynamics by inserting guitar solos on top (see Duarte 2020: 4). His solos give constant novelty to his music. They take the listener's attention and mark the musical passages. Along with these dynamics, his musical tempos are built without the support of automated metronomes. He says, "my biological timing is by itself a metronome". The tempo has to be mediated by emotions and musical characteristics. Thus, the pedal will be more than a programmed machine for programmed creations: it will be a piece of equipment that is open to non-programmable attributes.

However, Jorge doesn't ignore the need to respect fixed timings. He believes his performances have a natural tempo, provided by his experience playing with other musicians. For him, when artists play together, they naturally synchronise into an ideal timing for both. This synchronisation is relatively constant during the music execution. Being able to maintain a consistent tempo means having the ability to be respectful when playing with another person. On the other hand, this will expose those who cannot accompany a band and may socially exclude themselves. In Jorge's view, music is sociality, and making it is to be part of a community.

Because his originals don't contain lyrics, he considers that the music speed can relate to the message he wants to pass on to his listeners. A slow-tempo song will convey introspective and self-reflective feelings, whilst faster rhythms express festivity and sociability. Both introspection and sociability are positive practices that an artist can suggest to listeners. The artist can appeal to the synchrony between individual and social well-being and create harmony between both states. For Jorge, the levels of dynamics and tempos appeal to human care.

The relationship between Jorge and the loop pedal extends to stage performances. He considers that music arrangements for live performances are also a work of creativity. When using a loop pedal for concerts, Jorge's creative process is formed of the same three stages as his music for record production. First, he records a sound base formed by defined chords, which will be the guideline for placing other more dynamic sound elements. On the second recording track, he introduces rhythms of percussion derived from guitar percussion techniques, including slapping on the strings. He also often introduces sampled percussion through an electric organ containing recorded samples of drums and other percussions – adding the sample to the second audio track. Sometimes he has a bass guitar and tries to fit this with the percussion track. Finally, he adds variable guitar elements, such as solos, on the third audio track.

The musical introduction before the main body of the songs is essential for Jorge. Any music he has to create or collaborate on will allow him to make a musical entry that captivates people through suspense. In his way of structuring music by naming musical segments, this introduction comes before the first music part – part A. A specific loop can hold this section for the introduction. Further music parts and loops will be characterised by the counterpoint of guitars and intentional variations – pointing the difference and showing novelty to listeners (see Duarte 2020: 10). Introductions can also be made by a loop section created for added improvisations. This section for improvisation is planned to be a space for spontaneity. Usually, it consists only of the first recorded loop being reproduced and then complemented by the guitar's solo.

Improvisation is also an essential feature on stage when the loops created might come with a lousy start or ending while recording on the pedal. When these situations occur, Jorge says that, due to a concert's spontaneous nature, it is better to use the loop as it is recorded and correct it with another overlapping loop. If Jorge had to stop to re-record the first loop, the other artists playing would feel lost without the sound of Jorge's guitar. For this reason, he chooses to improvise without interrupting the repetition of the poor loop. He will instead add a newly recorded sound to complement or, in some cases, overshadow the first recording.

It is right to assume that music creation implies improvisation. Spontaneity may lead the artist to produce unexpected results that can contribute to the whole music production. It means that improvisation can be a conclusive idea and that the need to enhance the original concept will lead to more improvisations. The artist searches for possible avenues by improvising without fully knowing where to go. Still, they have memorised and skilled techniques to guarantee quality resolutions from the random space where improvisation occurs. These purposeful improvisations are actions looking for musical opportunities. Usually, after finding a sequence that pleases them, artists try to memorise and repeat it.

Jorge believes that it is necessary to be receptive to inspiration in order to achieve good results when improvising. He admits that inspiration does not always come, and he should not try to force it on a day when creative ideas aren't flowing. He believes preparing to take advantage of inspiration is the best method to create music. The rigorous training routine makes him more likely to take advantage of the periodic inspiration when it comes along. Being inspired but not being ready to manifest is regrettable. Also, it is unfortunate to be inspired one day but abandon that inspiration, hoping to recover it in the next few days. For Jorge, motivation will not come back quickly; if it comes back, it will be different.

There are also musical moments that serve as a bridge, which is a part that joins two essential music sections (Duarte 2020: 14). For example, according to Jorge, a song can have a bridge between parts A and B – the bridge is the specified and highlighted transition. Without any doubt, his music is built from essential fragments (see Girdzijauskienė 2015: 109): loop combinations, improvisations and bridges. Through constant repetitions, adjustments and resolutions, the artist reaches a whole that produces a unique and identifiable artistic work. In other words, these pieces of music are created with detailed and isolated elements.

Locations, identity, and loop pedal: conclusion

When viewing Jorge's online platforms, one can notice that some of the songs he performs are covers. As part of widespread knowledge, they are excellent vehicles for Jorge's relationship with his listeners. He explores covers as if they were his music and often devises an elaborate introduction to them. He sometimes does it without activating his recorder on the loop pedal; only after this phase will he start working with his feet on the pedal. When there is a repetitive layer of base chords, and enough time, the young man imitates the lead vocals with musical solos from his guitar. He does it by mixing some improvisation into it. He makes these guitar solos to hold the listener's attention quickly. It should be noted that whenever he transitions from low or medium notes to high notes on the guitar, something very emotional is awakened in those listening.

In the background, his loop pedal is also an instrument that mixes the various musical influences that Jorge carries with him. From Cape Verdean to American music, Morna to Blues, or whatever he may be asked to play, Jorge frequently uses his pedal to materialise the different genres he works with and mix these into his musicality. Jorge's music contains the hybridisation of everything he appreciates (see Kotarba and Vannini 2009: 136–137). Through his pedal, this mixture takes place, compelled by his experience and mastery of execution. The communion between Jorge's original views of traditional knowledge and modern opportunities from the loop pedal will dictate a unique musical and performative blend. This combination will ultimately be Jorge's statement and identity. "It's a difficult task to acquire a unique sound and essence so that they become an identity", he proudly admits.

As he learns and creates, Jorge comes across national and international artists who influence him a lot, like Voginha – with whom he played – Jimmy Dludlu[2] and Hernani Almeida.[3] Dludlu is a Mozambican singer who, having a connection with South Africa, acquired and promoted what is known as Afro-Jazz or Afro (Africa Interviews 2015), which mixes characteristics of American Jazz, Cuban Jazz, and several African genres. Afro Jazz became known on the continent arriving in Cape Verde through artists such as Hernani Almeida, who launched the Afronamim discographic work in 2009 (Hernanicv.net n.d.), immensely inspiring Jorge. Cape Verdean artists such as Hernani, including the very famous Tcheka and Mayra Andrade, have transformed national music into a world expression, and Jorge is part of this process. Afro-Jazz music became a professional development vehicle for Jorge, leading him to international interactions and performances in various music environments worldwide. Since 2010 Jorge has played with notable national and international artists such as Ana Moura, Dino D'santiago, Djode from Broda Music, Fattú Djakité, and Hilario Silva.

Despite all the collaborations, Jorge hopes to record a music album of his instrumentals. Thus far, he has only been in studios to record guitars for other artists and design some of his music outside his loop pedal. Jorge is also part of Azagua, a musical group comprised primarily of young and talented artists on the island of Santiago. The music he mostly wants to produce is

Afro-Jazz. Its syntax has the expression "afro", which comes from "Africa" or "African". This identification is not unreasonable. Africans are adopting Jazz and giving it new dynamics. It could be said that Africans rescued Jazz: this genre was created by African descendants living in the US; therefore, Afro-Jazz represents the expected junction of 'diverse' African people separated by the ocean (see Hall 1995: 9). Like many other African artists, Jorge claims Jazz in the same way he promotes African identity in his works.

This desire for meaning and cultural manifestation comes from the Cape Verdean history itself – a Portuguese colony until a few decades ago – influencing the imagination and music of young people like Jorge (see Turino 2010: 182). We speak of people who are more politically critical, questioning social problems and condemning the injustice of neo-colonialism and capitalism in the modern world (see Fanon et al. 2004: 9–12). Considering Jorge's standpoint as an African (see Harding 2004: 8), the Afro-Jazz concept is also to shake up the political and economic differences in the relationship between the West and the Global South countries. Jorge personally seeks these statements through musical mastery and aesthetics. This unique musicality has loop equipment at its centre. The loop pedal isn't only a musical aid but a form of music production for new cultural construction, which re-places Cape Verdean people in local and global spaces. Loop pedals are also instruments that facilitate both personal and social meditation and well-being. These new pieces of equipment achieve this transversal role through constant support for an artist like Jorge by shaping the contraposition of his actions and intentions regarding cultural and individual expectations. Jorge drives his musicality and social, cultural and political orientation through a loop pedal by promoting constant experimentation, improvisation, and resolution.

Notes

1 "Ian Ethan Case - Open Land Music." n.d. Ianethan.com. https://ianethan.com/.
2 "Jimmy Dludlu." n.d. Africa Interviews. http://www.africainterviews.com/jimmy-dludlu/.
3 "Hernani Almeida." n.d. Hernanicv.net. http://hernanicv.net/menu/about_hernani.htm.

References

Duarte, Alexsander. "O Uso Da Loop Station Em Performance Musical: Implicações E Exigências Interpretativas." *Revista Vórtex* 8, no. 2 (2020): 1–3A.
Fanon, Frantz, Constance Farrington, & Jean-Paul Sartre. *The Wretched of the Earth*. (London: Penguin Books, 2004).
Girdzijauskienė, Rūta. "How Pupils Create Compositions: The Analysis of the Process of Music Creation." *Problems in Music Pedagogy* 14, no. 1–2 (2015): 107–18.
Hall, Stuart. "Negotiating Caribbean Identities." *New Left Review*, no. 209 (1995): 3–14.

Harding, Sandra (ed) *The Feminist Standpoint Theory Reader: Intellectual and Political Controversies*. (New York and London: Routledge, 2004).
Kotarba, Joseph, & Phillip Vannini. *Understanding Society through Popular Music* (New York: Routledge, 2009).
Schloss, Joseph Glenn & American Council of Learned Societies. *Making Beats the Art of Sample-based Hip-hop*. (Middletown, CT: Wesleyan University Press, 2014).
Turino, Thomas. "The Mbira, Worldbeat, and the International Imagination." *World of Music* 51, no. 1–3 (2010): 171–92.

Links

Jorge: https://www.instagram.com/tv/CNQcK-LDKvY/
Jorge and Ana Moura: https://www.youtube.com/watch?v=RAQ05akmD-w
Jorge and Nelson Freitas: https://www.instagram.com/tv/CUrnXGgICz2/
Jorge and Azagua: https://www.youtube.com/watch?v=ZwWIb8c6RwI
Ian Case: https://www.youtube.com/watch?v=wa4wLz—EdY
Voginha: https://www.youtube.com/watch?v=os77zkOnBlQ
Jimmy Dludlu: https://www.youtube.com/watch?v=e1Y_rVb7n-w
Hernani Almeida: https://www.youtube.com/watch?v=WDPV9vXIH7M
Tcheka: https://www.youtube.com/watch?v=393yuRmZeUo
Mayra Andrade: https://www.youtube.com/watch?v=R5p0MI9BtFk

10
TAKING THE LIVE OUT OF LOOPING – COMPOSING WITH THE LOOP PEDAL

Aoife Hiney and Isabel Novella

Singer-songwriter. What image or images does that conjure up for you? And a loop pedal? Maybe the singer-songwriter image is that which Nicholas Cook describes as the frequently portrayed "composer-cum-lyricist struggling to express his or her innermost feelings" (Cook 2018: 38). Or perhaps it aligns with Pamela Burnard's introduction to the singer-songwriters featured in her book on musical creativities, in explaining that the term singer-songwriter refers to "artists who both write and perform their material and who are able to perform solo, usually on acoustic guitar or piano" (Burnard 2012: 75). As Bennett explains, the term "singer-songwriter" is associated with a set of cultural assumptions which involve both the artist and their creative output, namely that "songs will be implicitly autobiographical, performed on piano or guitar, and expressive of the writer's own thoughts, feelings and worldview" (Bennett 2015: 43). As for the loop pedal, it's quite likely that it conjures up images of specific performers, such as Imogen Heap or Ed Sheeran, or a general idea of a performance in which the artist is recording, playing and overdubbing loops of themselves while performing (Knowles & Hewitt 2012). However, this chapter will explore a different facet of the loop pedal, an approach that does not necessarily involve live performance.

This text is the result of a collaboration between Isabel Novella and Aoife Hiney. Isabel Novella is a singer, songwriter and composer from Mozambique. She has been living in Lisbon, Portugal, since 2017, while pursuing a degree in Music Production at the Escola de Tecnologias, Inovação e Criação in Lisbon. Isabel has released two albums, *Isabel Novella* (2012) and *Metamorfose* (2019) and has performed widely in Mozambique, South Africa, the USA, and various European countries. Aoife Hiney is a choral conductor, teacher and researcher. She is from Ireland but has been based in Aveiro, Portugal since 2010.

We began our collaboration with an initial interview via Zoom, which we recorded and Aoife later transcribed and translated from Portuguese to English. This resulted in an initial text, which we then emailed back and forth to each other until we were happy that all the information was correct. However, we felt that something was not right about the text, and we decided not to submit it to this book's editors at that point. Although all the information had been produced in a collaborative capacity, the actual writing of the text was – initially – mostly Aoife's work. The problem was that our positions within that text were abundantly clear: Aoife was the researcher and Isabel the subject, which reflected neither the reality nor the aims of our collaboration. We felt that one of the main issues was the way in which our writing was leaning on excerpts from our first conversation: they read too much like 'collected' data as opposed to data that was collectively produced.

We subsequently had a long conversation about our options – both of us felt that the text would be richer if Isabel spoke about her experience in the first person. However, if both Aoife and Isabel were "I", the text would be too confusing for the reader. This question of voice and representation is by no means new (Denzin & Lincoln 2011), but the question of collaboratively writing a text that focuses on the personal experiences of only one of the writers brings certain challenges, particularly when contextualising these experiences with literature, for example, which Hellier describes as the difficulties surrounding the creation of a narrative that "balances the tension between ideas and people" (Hellier 2013: 9). We have opted to 'zoom in' and 'zoom out', with Isabel sharing her experiences in her own words, contrasting with sections that aim to demonstrate the wider implications of those experiences.

At this point, we should explain a little about our first meeting. Aoife had listened to some of Isabel's music and came to the meeting expecting to learn about how Isabel uses the loop pedal in performance contexts. Isabel described herself as being fascinated by the loop pedal when she saw one of her brothers performing on guitar with a loop pedal. She then explained that she began to question whether, as a singer, she could also make music with a loop pedal. Isabel's experiences with the loop pedal began in 2011, and by 2012 she was using the loop pedal in live performances and in the production of her first album.

Hoad and Wilson describe live looping as "the recording and playback of a piece of music in real time using loop pedals or laptop-based audio interfaces" (2022: 84). Similarly, Knowles and Hewitt state that the loop pedal allows the performer to "record, play and overdub loops of themselves while performing" (2012: 13). Pamela Hulme (2020), a loop artist and pipe organist, explains that her research demonstrates that across genres, there is "commonality of approach" in relation to live looping practice. Meanwhile,

Keens describes the processes of recording, improvising, and performing in 'real time' (2019: 13). In relation to the possibilities of the loop pedal in live performance contexts, Isabel explains that in 2011, when she began experimenting with the loop pedal, she was frequently performing acoustic sets. She began to wonder if she might be able to perform alone on stage, just her and the loop pedal, and began to use the loop pedal for some acoustic concerts in Mozambique. However, Isabel regularly performs with a band, featuring keyboard, bass guitar, lead guitar, drums, and percussion. Although she has experimented with using a loop pedal together with the band, she felt that it was too much of a risk, and it presented various problems in terms of synchronisation. As Duarte (2020) states, incorporating the loop pedal in performance is not without its challenges in terms of coordination and synchronicity. Waite (2018) also describes the difficulties surrounding performing using a loop pedal to recreate a recording of an album, primarily citing the challenges of simultaneously singing, playing guitar, and operating controllers. Although Isabel feels that some combinations of instruments can work well together in the context of live looping performances, in general, in Isabel's performance practice, the loop pedal is principally for solo and acoustic sets.

At that point, our conversation rapidly turned to how Isabel uses the loop pedal in her compositional processes, bringing a new element to the conversation. It is the impact that the loop pedal has had on her work as a composer and songwriter that is the most striking, bridging composer-performer collaboration and the potential of the loop-station for singer-songwriters. Although there are some studies which consider the loop pedal as a compositional device, the literature has largely focused on the loop pedal in live performance. Yang provides a broader perspective of the potential of the loop pedal, illustrating his point by describing how the composer Philip Sheppard "prefers writing from behind his cello" (2022: 55) through a loop pedal, and concluding that "live looping can serve as a device for experimentation during the composition process while also allowing the musician to create a finished work" (2022: 55). However, Hulme suggests that despite its improvisatory and spontaneous nature, using a loop pedal in performance is based on "a hidden complexity that requires pre-determination, in other words: composition" (2020: 43). Thus, this chapter examines how employing a loop pedal during the compositional process can contribute towards the development of autonomy, particularly for composers that tend not to work with traditional music notation but prefer to use oral approaches.

Towards composition

> I grew up in Maputo, Mozambique, surrounded by music. I am one of eight children; my father and all my siblings also participate in musical

activities. I began singing in a choir at the age of five. By the time I was fourteen, I was active as a backing singer for various bands, performing both on stage and in the studio. This experience inspired me to pursue a career in music, while the opportunity to perform a wide range of musical styles, from hip-hop and Marrabenta to Western art music was an education in itself. I had singing lessons focusing on jazz and Brazilian music while visiting The Netherlands. This melting pot of performance experiences and lessons leads me to describe my musical style as a balance between African rhythms and Western influences.

I used to always write my lyrics and my ideas in notebooks or on pieces of paper. Thankfully, I have a good memory, and in Mozambique I had my brothers at home and friends in the studio – instrumentalists, pianists, guitarists and beat makers – so I would literally run there and say "I have this melody, can you play it?" I was lucky that some people would stay with me until we reached exactly what I had been thinking. I always had, and have, everything in my head, but sometimes it is difficult to explain, so it is hard to say "no, not that note, this one". I can sing the melodies of the guitar, piano, or drum parts, and sing the chords, but I can't play the instruments.

Impett describes composition as "an activity simultaneously of aural and aesthetic imagining, in some kind of feedback relationship" (2009: 408). This view corresponds to Isabel's experiences, as she imagines the sounds that will accompany her song melodies, and in terms of aesthetic imagining, she explains that she knows instinctively what harmonies will work well. We have referred to this in our conversations as tacit knowledge that was probably acquired through performing backing vocals, as Isabel says she cannot explain how she knows this or how she learned what each voice would need to sing at any given moment. Similarly, Isabel says that she cannot explain what the sounds should be by using words or any graphical representation – but she can hear them and sing them. Isabel studied piano and music theory as a child, but she does not use traditional musical notation to register her melodic ideas, and as Bennett (2015) and Tobias (2013) observe, the teaching and learning outcomes relating to a traditional musical education in terms of musical literacy and aural skills frequently do not develop the necessary skills for songwriting.

Although Cook (2018) states that the role of notation is not usually significant for rock and pop music, while Bennet (2015) explains that it is the audio itself that is the text in the context of popular music, Isabel feels that not having the skills to demonstrate exactly what she wanted to hear on a given instrument frequently hindered the rehearsal process when collaborating with other musicians. Bennett (2012) describes songwriting as a primarily aural experience, with the melody either memorised or recorded, with lead sheets to register the lyrics and guitar chords. However, Cook, in exploring blogs and resources dedicated to songwriters relating to imagination and creativity, notes that "you need some kind of technical understanding to know what to do with the ideas that come to you" (2018: 73), which resonates with recommendations for 'book learning' advocated by one of the bloggers featured in Cook's text. Similarly, both singer-songwriters interviewed by Pamela Burnard (2012) in her exploration of musical creativities were initially trained in the Western classical tradition, well grounded in traditional musical notation and harmony, for example. As an adult, Isabel experimented with guitar and piano lessons as a means of developing another channel to express her musical ideas. As Cook explains, musicians "learn to think with instruments in their hands or mouths, and so the material affordances of instruments equally condition imagination" (2018: 133). However, Isabel found that singing and playing simultaneously was challenging, and that needing to concentrate on the guitar or the piano hampered her vocal creativity.

> I started writing when I was very young, when I was working as a backing singer for various groups over the years. I always wrote my own songs, but the difficulty I had was not being able to compose something from start to finish, and not being able to hand it to someone and say "right, this is what I want". My first contact with a loop pedal was through my brother. Now with the loop pedal, the process works really well, because I can just sing everything and then say to the instrumentalists "OK, this is what I want to hear, or I'd like it to be like this", and it makes everything so much easier.

Collaboration is, of course, a well-established practice, particularly for the writing of commercial music, and although Cook (2018) mentions that collaborations are rarely demarcated, Bennett explains that successful songwriters are often "'topliners'" – writing only the melody and the lyrics, while the producers "provide backing tracks and post-production editing" (2015: 54). The literature surrounding songwriting demonstrates the importance of production, and the skills surrounding music production. Knowles and Hewitt argue that "recording and performance practices are trending towards each other and that this is underpinned by technological shifts, a change in the level of production literacy of musicians broadly, and an increasing shift toward more technologically intensive performance" (2012: 1). Bennett identifies seven creative contributions that the author terms 'track imperatives', (2015: 45) explaining that songwriters not only need to understand these 'track imperatives' but also know how they should be balanced, namely production, instrumental performance, vocal performance, arrangement, melody, lyrics, and harmony. Thus, Tobias (2013) argues that aural skill training should include sonic elements of a recording, from spatialisation and equalisation to effects processing. However, both Pamela Burnard (2012) and Katelyn Barney (2007) point to a gender imbalance in music production. Burnard comments on how very few women work in music production and technology (2012: 78), while Barney describes how very few women were enrolled in a studio music course she took, and that the male students "seemed to have more background in using technology and dominated class workshops and activities" (2007: 109). Although neither of these texts is particularly recent, Isabel's experience largely mirrors the reality described by Burnard (2012) and Barney (2007), with only three other women in her class when she studied music production, but approximately twelve men.

Composing with the loop pedal

> In composing with the loop pedal, you discover all these possibilities, ways of playing with your voice, in various registers, in different tonalities, trying new sounds.
>
> I worked on the new single with my brother, Isildo. I was humming and recording it on the loop pedal. I did the vocals and percussion lines, and then decided to do the idea for the bass line as well, before I sent it to my brother . . . But I knew what I wanted to hear, because I was using a rhythm from the north of Mozambique, called Tufo. It's music and dance performed with specific drums, with a distinct rhythm, and I had the privilege of experiencing it in Nampula, and I wanted that sound. So, my brother said "look, you're explaining, but I'm not following. Why don't you do it? Sing what it is you want using the loop pedal, and that way I'll understand". So, I did, and sent it, and he said "OK!", and from then on, we were on the right path, and the song is really beautiful and surprising.

Marrington (2017) considers that the Digital Audio Workstation (DAW) is an instrument in its own right, and that its place in the compositional process is akin to the role of the guitar or piano in what the author describes as 'previous eras' of songwriting. However, the use of a DAW in songwriting and composition has frequently focused on rearranging and reimagining sound samples, while Isabel's approach is centred entirely on her own original sounds. The use of the Loop Station during the compositional process has also facilitated the inclusion of Isabel's African influences. As DeNora explains, "Musical materials provide terms and templates for elaborating self-identity – for identity's identification" (2006: 145). Impett explains that the evolution of recording technologies has meant that instead of being a means of documenting a performance, these technologies have contributed to the development of a new situation, "in which compositional exploration may be audibly inscribed" (2009: 409). Thus, the loop pedal allows Isabel to hear her various melodic and rhythmic ideas simultaneously, and, crucially, to be able to communicate her ideas to other musicians. This oral approach is similar to the rehearsal processes promoted by the American jazz musician Charles Mingus. Toynbee (2006) describes how Mingus would rehearse

> his musicians without using a score (. . .) he would introduce them to the band by singing or playing the melody line for each instrument. Musicians had to learn their part by ear without even a chord chart to refer to. In an important sense, composition, improvisation, and recording are elided in the Mingus method.
>
> *(Toynbee 2006: 73–74)*

Indeed, in Isabel's experience, too much weight can be given to the development of theoretical knowledge and a reliance on traditionally notated music, to the detriment of creativity and informal communal music making.

> I often went to jam sessions in The Netherlands, where everyone would have a score in front of them. And I would think to myself, "I don't want to go to a music school, because I don't want to become 'square', too focused on everything being 'correct'.' But it's a question of studying, of seeking knowledge, but you also have to put your heart and soul into what you're doing. Otherwise, we're all just going to play 'do, re, mi', and nothing will happen, because school gives us all the tools we need to make good music, but then each individual has to bring something unique to it, from our hearts, and from our souls".

Similarly, Théberge considers that style is "something that is primarily felt, it is an awareness that is as much physical as it is cognitive (. . .) something acquired only through an extended process of learning through practice" (2006: 286). Thus, although the Loop Station allowed Isabel to exert more control over her compositions, and to communicate her ideas clearly, she still sees her performance practice as a collaborative process, as individual performers further develop her initial ideas.

> I was performing in South Africa, and it was the first time I was working with these musicians and their response was "ok, great. Now we know what you want, we can go from there". Because I would never say "this is my idea and that's it, you have to play it this way". I prefer to say something like "this is more or less what I want, but of course you know your instrument better than I do, but I'd like to hear this sound, or this melody".

Isabel's view thus resonates with Philip's (2004) perspective on the reciprocal process involved between the performer and the composer, with the composer learning from and being influenced by the performers.

Creating *Oração*

Isabel launched her single *Oração* ("Prayer") in May 2021, written in response to the Coronavirus (COVID-19) global pandemic and conflict, particularly in Cabo Delgado. Composed by Isabel Novella, it was produced by Isildo Novela and features Preben Carlsen (guitar), Tobias Lautrup (violoncello), and Sylvester Agbedoglo (percussion). In addition to playing electric bass, Isildo Novela also added synthesised violin and drum programming. The track was edited and premixed by Isildo Novela at Mbilo Studio in Copenhagen, with the final mixing and mastering by Ossian Ryner, based in Malmo, Sweden.

Isabel explains that she had recorded fragments of the melody on her phone and had started putting some harmonies together using the loop pedal, but had not yet given much thought to the lyrics beyond some initial ideas. However, an opportunity to perform on a television show was the impetus for her to advance very quickly in order to have a new song ready in response to the violence in northern Mozambique.

Isabel's inclusion of Tufo rhythms in the song, and her wish to include an authentic Tufo group in the video for *Oração*, was a chance to honour that tradition but also to speak about what was happening in her country. Tufo is a religious singing and dancing practice in praise of the prophet Muhammed and accompanied by four flat drums or tambourines (Arnfred, 2004). Tufo was originally danced by men during certain Islamic celebrations, but is now danced by women, and competitive Tufo dance groups have also emerged in the northern, coastal region of Mozambique (Hebden 2020). Similarly, Arnfred explains that "women's Tufo groups are an expanding culture; new groups are mushrooming all along the coast and as far inland as Nampula city, the largest provincial capital in the north" (2004: 40). In addition to its geographical expansion, Tufo groups are now also present in other contexts aside from the Muslim tradition, and song lyrics may now include contemporary and/or political themes apart from the religious repertoire.

> The melody of the introduction came to me, and I recorded it on my phone. Then I started putting ideas down using the loop pedal. I didn't have the words yet, I wasn't sure what it meant or what it would mean, but I often record my ideas and then I'll find the words that match the music, that mirror the feeling of the music, words that make sense.
>
> At the time, I had been invited to go on television, and I wanted to perform this song on television because I don't like to go on television without having something to say. I had already given the song its title, and I had the chorus, but I still didn't have a verse. But wanting to have Oração ready for the television appearance wasn't just to have a new song, it was to have an opportunity to talk about what was happening in Mozambique.

When I was seventeen years old, I went to northern Mozambique as a volunteer. I taught English to adults, and I started a cultural centre for kids and women where we would teach them singing and dance, and I had the opportunity to learn from them as well. And so, I wanted to include the Tufo rhythm in the song. I was trying to explain it to my brother, but he's not that familiar with Tufo although he has seen the dance. He said to me "you know exactly what you want, record it and send it to me". Then I recorded the percussion and the bass line idea and the vocal harmonies using the loop pedal.

About the text, I often mix different languages together. I think that words themselves have their own musicality. For Oração, I remembered the story from the Bible, of Pentecost, and the flames of fire reaching down, and the disciples began speaking in tongues. For that reason, Oração has a mixture of three languages – Hebrew, because the majority of the population in the north of Mozambique is Muslim, Portuguese, and Changana (the most predominant language in southern Mozambique).

Oração is a prayer, it doesn't have sentences as such, but phrases. It's about what we are feeling, it's the Holy Spirit in me, but for people who don't believe, it can be something else. The word "moya", for example, can mean soul, or Holy Spirit, or air. So, it can be any of those things.

When I was writing the lyrics, I just thought "violins". I was talking to my brother (bass player and producer Isildo Novela), and I told him that I was thinking of adding strings to a certain part and demonstrated what I was imagining. He just said, "wait a moment", and he sent me a file where he had already added a violin part, exactly what I wanted. We have some kind of telepathy!

I sang Oração on television, and then recorded it in the studio. It's special because we don't just sing, we send messages. We don't always get opportunities, but when these opportunities come – like television appearances – we have to use them to speak, to use our voices and to send a message. And very often we talk about what has already happened, it's rare that we get a chance to talk about what is actually happening right now.

The video for Oração was made in Mozambique and in Portugal. I wanted to have both places included. I know a Tufo group in Maputo, and I wanted to involve them. I wanted to have this dance represented, but by people who are from that culture. I contracted a company in Mozambique to film the Tufo group, they sent me their footage and then I did my part in Lisbon.

The loop pedal and agency

The question of agency in relation to the loop pedal is most frequently related to how the pedal can replace a larger ensemble, effectively allowing an artist to perform solo (Hoad & Wilson 2022; Keens, 2019; Knowles & Hewitt, 2012). Hoad and Wilson conclude that "for all our interviewees, looping was a mode of declaring self-sufficiency in the face of societal, capitalist hegemony" (2022: 93). However, if we consider the link between composition and musical autonomy coupled with the role of production, as according to Tobias, "songwriting and composing can thus be seen as overarching processes that include production" (Kaschub & Smith, 2013: 214), musical agency can thus encompass both composition and production.

Isabel describes her discovery of the loop pedal as a compositional tool as bringing her freedom and developing her autonomy as a composer. Furthermore, the difficulties that she had felt due to not playing an instrument were overcome due to the possibilities afforded by the Loop Station. Kaschub and Smith (2013) explain that composers have always faced the challenge of transferring their musical ideas for preservation and potential sharing. However, the authors suggest that these challenges are now not so unsurmountable.

> Access, compositional context, and the very definition of who can be a composer all have been expanded by technological means. Current ways of crafting and controlling musical experience provide a freedom of time and place that allow people to focus on their own musical pursuits and their individual creativity. (. . .) The challenges of notating and preserving musical ideas (. . .) have been substantially reduced by the advent of highly accessible tools.
> *(Kaschub & Smith 2013: 5)*

Similarly Savage and Challis, in their study of a music composition project, state that using music technology in composition allowed the students a means of expressing their musical ideas without requiring a 'traditional' instrumental education and the development of the musical skills associated with that approach (2001: 146). Echoing this, Keens (2019) explains that musical creation is now accessible to those from trained and non-trained backgrounds and mentions that this technology is rapidly evolving and user-friendly. Isabel describes the impact of the loop pedal on her composition:

> I started to feel a sense of liberty, being able to write my songs without having to depend on anyone else. It began just as sheer curiosity, but the Loop Station has given me such freedom, so that now I can always write my own songs. As a singer without another instrument, composing was difficult, but now I can sing everything on my own, as the Loop Station gives me all this freedom, it's absolutely fascinating.

This autonomy, which has promoted greater confidence in her musical abilities and ease of communication, dovetails with Isabel's study of music production. She describes the study of music production as another means of developing her autonomy, and strongly advises other musicians to experiment with the Loop Station in their musical creation.

> I am a producer, I can produce my own music, I can compose my own songs from beginning to end just with my voice, and I see the Loop Station as a new tool for composition and production, because not all musicians can play instruments, so I tell everyone to experiment and try to use the Loop Station. I have some friends who perform Hip-Hop in Mozambique, and I've told them "try this, you'll have more freedom", or it opens other possibilities, because often, with Hip-Hop, they know what sound they want to hear, or how they want it to sound, but the other musicians don't always get it, so it's a tool, an extension of our voices.

In relation to production, Knowles and Hewitt (2012) point to an increase in the musician's production literacy. Barney's research featuring four Indigenous Australian women performers focuses strongly on the question of agency and the importance of their individual artistic expression during the recording process. The author explains that "a resounding theme in their comments is that by being independent artists Indigenous women can sustain control and power over the recording process" (2007: 110). In fact, perhaps one of the most striking excerpts of Barney's interview with the performer Lexine Solomon is the statement "only I know what I am looking for – I become the executive producer of my work" (Solomon in Barney 2007).

Closing comments

Isabel's journey with the loop pedal has led to the development of her performance practice, but the greatest impact has been on her compositional process. Isabel, through her composing with the Loop Station, has developed creative autonomy in her compositions, while her experience in music production has led her to develop skills to contribute towards technical autonomy as well, and she views music production as highly creative.

> I decided that I wanted to be involved in the most creative part, because I like to write, I am always writing or singing a melody, composing something, and I said "I want to do this creative part, music production, for me and for Mozambique". What I really want to do is finish my course and be able to share some of what I have learned, especially because I am a woman. As far as I know, at the moment we do not have women involved in production in Mozambique, although we have many female singers and instrumentalists, and I think it is a really beautiful part of music making that we need to explore.

Isabel also spoke about how very often singers arrive at the studio when the instrumental has already been finalised by the producer. This limits the singer's involvement in the creative process, as their role is thus restricted to simply adding in their melody, with very little room for modification at that point. Similarly, Barney's (2007) interviews with women performers in Australia, for example, depicted a number of issues they faced in the recording studio, and the difficulties surrounding maintaining their control of the artistic process.

Although we are reluctant to generalise, our conversations have led us to some tentative conclusions, particularly in terms of the potential of the loop pedal as a compositional tool. Hallam, Cross, and Thaut (2011) explain that specific conventions and rules for processes of composition and improvisation are associated with particular musical contexts. Similarly, Burnard (2012) criticises the myths surrounding creativity and compositional processes, while Théberge (2006) describes the increasingly prescriptive role of musical notation. These conditions can prove exclusive or daunting for people that may not have had formal musical instruction, who have not developed the specific vocabulary that surrounds these conventions, or a working knowledge of the fundamentals of music theory that this vocabulary represents. Martin (2012) explains that there is a need to move away from using technology for more traditional composition techniques (such as notation and notation-based midi input), and that teaching and learning activities should reflect a diversity of approaches to composition. Similarly, Bennett (2015) explains that students should be introduced to the ways of being musical though recording, engineering, and mixing processes.

The most striking aspect of these possibilities can perhaps be summarised as the development of musical agency. Tobias (2013) in their text regarding approaches to composing, songwriting and producing in the context of a high school in the USA, highlights the role of technology in helping students to gain and sustain agency, and the importance of being able to speak with producers and engineers, or to be able to communicate with an artist if they themselves are in that role. We feel that the loop pedal has the potential to be a powerful tool facilitating musicians to develop their musical agency.

References

Arnfred, Signe. "Tufo Dancing: Muslim Women's Culture in Northern Mozambique". *Lusotopie* 11 (2004): 39–65.
Barney, Katelyn. "Sending a message: How indigenous Australian women use contemporary music recording technologies to provide a space for agency, viewpoints and agendas". *The world of music* (2007): 105–124.
Bennett, Joe. "Creativities in popular songwriting curricula: Teaching or learning". In P. Burnard & E. Haddon (Eds.). *Activating diverse musical creativities: Teaching and learning in higher music education* (London: Bloomsbury, 2015), 37–56.

Bennett, Joe. Constraint, Collaboration and Creativity in Popular Songwriting Teams. In *The Act of Music Composition – Studies in the Creative Process*, edited by Dave Collins (London: Routledge, 2012).

Burnard, Pamela. *Musical creativities in practice*. (Oxford: Oxford University Press, 2012).

Cook, Nicholas. *Music as creative practice* (New York: Oxford University Press, 2018).

DeNora, Tia. Music and emotion in real time. In *Consuming Music Together* (Springer, Dordrecht, 2006), 19–33.

Denzin, Norman & Lincoln, Yvonna (Eds.). *The Sage Handbook of Qualitative Research* (London: Sage, 2011).

Duarte, Alexsander. "O uso da Loop Station em performance musical: implicações e exigências interpretativas". *Revista Vórtex* 8, no. 2 (2020).

Hallam, Susan, Cross, Ian, & Thaut, Michael (Eds.). *Oxford handbook of music psychology* (New York: Oxford University Press, 2011).

Hellier Ruth. (Ed.). *Women Singers in Global Contexts. Music, Biography, Identity* (Urbana: University of Illinois Press, 2013).

Hebden, Ellen. "Compromising beauties: affective movement and gendered (im)mobilities in women's competitive *tufo* dancing in Northern Mozambique". *Culture, Theory and Critique* 61, no. 2–3 (2020): 208–228, DOI: 10.1080/14735784.2020.1858127.

Hoad, Catherine & Wilson, Oli. "Looping Alone, Together. Music, community and environmental self-sustainability in Aotearoa/New Zealand", in *Mixing Pop and Politics. Political Dimensions of Popular Music in the 21st Century*, edited by Catherine Hoad, Geoff Stahl & Oli Wilson (London: Routledge, 2022).

Hulme, Pamela. "Manipulating Musical Surface: Perception as compositional material in live looping and organ with electronics" (Master's thesis, University of Huddersfield, 2020).
http://eprints.hud.ac.uk/id/eprint/35280/1/FINAL%20THESIS%20-%20Hulme.pdf

Impett, Jonathan. "Making a mark". In *The Oxford handbook of music psychology* (USA: Oxford University Press, 2009), 403.

Kaschub, Michele, & Smith, Janice. (Eds.) *Composing Our Future. Preparing Music Educators to Teach Composition* (Oxford: Oxford University Press, 2013).

Keens, Heather Grace. "The 'Autonomized Performer': Operatic Voice and Looping". (Master's Thesis. Mcquarie University, 2019).

Knowles, Julian & Hewitt, Donna. "Performance recordivity: studio music in a live context". In Burgess, Richard James & Isakoff, Katia (Eds.) *7th Art of Record Production Conference*, 2–4 December 2011, San Francisco State University, San Francisco, CA, 2012.
https://eprints.qut.edu.au/48489/1/Knowles_J_Hewitt_D_ARP2011.pdf

Marrington, Mark. "Composing with the digital audio workstation". In *The singer-songwriter handbook*, edited by J. Williams & K. Williams (London: Bloomsbury, 2017), 77–89.

Martin, Jeffrey. "Toward authentic electronic music in the curriculum: Connecting teaching to current compositional practices". *International Journal of Music Education 30*, no. 2 (2012): 120–132.

Philip, Robert. *Performing music in the age of recording* (Bury St Edmunds: Yale University Press, 2004).

Savage, Jonathan & Challis, Mike. "Dunwich revisited: Collaborative composition and performance with new technologies". *British Journal of Music Education 18*, no. 2 (2001): 139–149.

Toynbee, Jason. "Making up and showing off: what musicians do". In *The popular music studies reader.*, edited by A. Bennett, B. Shank, & J. Toynbee (London: Routledge, 2006).

Théberge, Paul. "Music/Technology/Practice: Musical Knowledge in Action". *The Popular Music Studies Reader* (London and New York: Routledge, 2006).

Tobias, Evan. "Composing, songwriting, and producing: Informing popular music pedagogy". *Research Studies in Music Education, 35*, no. 2 (2013): 213–237.

Waite, Simon. "Networks of Liveness in Singer-Songwriting: A practice-based enquiry into developing audio-visual interactive systems and creative strategies for composition and performance" (Doctoral Thesis. De Montfort University Leicester, 2018).

Yang, Caleb. "Creative Practice for Classical String Players with Live Looping" (Music Theses, 2022). https://repository.belmont.edu/music_theses/7.

Internet resources

https://www.youtube.com/watch?v=eMuIVW_G0qQ
https://mbenga.co.mz/blog/2021/05/21/oracao-de-isabel-novella-por-cabo-delgado/

11
THE ONE AND THE MANY

Interview with Portuguese singer and songwriter Joana Lisboa

Samuel Peruzzolo Vieira and Joana Lisboa

(Non-musical) prologue

A central figure in Portuguese literature, Fernando Pessoa (1888–1935) dwelled in his many heteronyms.[1] During his life, Pessoa mastered the craft of bringing to life numerous poets, having not only conceived each one, but giving life to and living through them. An immeasurable palette of made-up biographies and personalities can be found throughout his vast work. Among the many, Alberto Caeiro, Álvaro de Campos, Bernardo Soares, and Ricardo Reis are some of the most well-known heteronyms Fernando Pessoa created.

The fact of personifying dozens[2] of imaginary poets is remarkable in itself, but what makes Pessoa exceptional is the vividness and distinctiveness he gave to each heteronym. Engaging with those invented writers might lead the reader to the uneasiness of establishing a relationship of trust with the author yet knowing that they are, nevertheless, fictional. Alberto Caeiro is a product of Pessoa's imagination, a conceptual manifestation of someone else's mind. But despite this deceptive revelation, Caeiro's works are as cohesive and profound as they can be: one can easily recognise his authorship.

In a letter to another Portuguese writer, Fernando Pessoa brilliantly describes the process of conceiving heteronyms:

> I see before me, in the colourless but real space of the dream, the faces, the gestures of Caeiro, Ricardo Reis and Álvaro de Campos. I build their ages and lives. I then created a non-existent *coterie*. I fixed it all in moulds of reality. I graduated the influences, knew the friendships, heard, inside me, the discussions and the divergences of criteria, and in all of this it seems to

me that I was the creator of everything, the least that was there. It seems that everything happened independently of me.

(PESSOA 1999: 334)

Far from being schizophrenic, Fernando Pessoa is a great example of one of the most complex novelists of the twentieth century. In the entirety of his work lies the greatness of a writer dedicated to elaborating and claiming his singularity. Pessoa not only envisioned and designed his diverseness in infinite detail, but confronted and gave voice to the multiple personas that lived inside him. Fascinating voices they were, as prolific as can be.

At this point, it seems suitable to shift the discussion towards music, for the analogy between heteronyms and live looping is evident. Enhanced by the possibility of examining the echoes of our inner voices, activities of poetic writing and music-related looping suggest an accentuation of the practices of self-analysis, communicability, structural thinking, project management, and interdependency between parts and the whole, much alike in both cases.

As a matter of fact, the mention of Fernando Pessoa is not by chance. The featured artist in this chapter, Portuguese singer and songwriter Joana Lisboa (1983), cites the famous Portuguese writer during the interview. Through her own words: "I identify myself with Pessoa's worlds, due to their existential and modernist side, however charmingly simple. I also identify with his heteronym Alberto Caeiro, for the way he embraces nature inside himself" (Lisboa 2021).

Musical inspirations

Joana Lisboa is the author of the project entitled *Uma música por dia nem sabes o bem que te fazia*, which in English would mean something like "A song a day makes you feel okay". The project took place in the summer of 2013 and consisted of a 33-day immersion in a compositional and performative state of flow in which a new song had to be composed, performed and recorded on a daily basis. This experience gave birth to her EP entitled *Rascunho*[3] (Lisboa 2013).

According to Lisboa, the key factor for developing the project "*Uma música por dia nem sabes o bem que te fazia*" lays both on the autonomy and on the exploratory environment provided by the loop pedal (LP). Regarding the former, apart from two specific exceptions, only her voice was used as a musical instrument in the whole project. Lisboa justifies the choice for such a challenging project by reporting that its goal was "to prove to myself [Joana] that I was sufficiently capable and that I could make everything by myself" (Lisboa 2020).[4] Initially, the idea was to form a band which created their own songs, but as she encountered difficulties in finding available and willing musicians, she opted to work alone. As she had not yet mastered any harmonic instrument at that time, she "decided to create everything with the instrument that [she] dominated: [her] voice". It is not by chance that the above constitutes an important aspect of this research and provides the basis for some preliminary assumptions, as follows:

(1) contrary to expectations, the fact of being unable to accompany herself with another instrument would not be less favourable when choosing a live looping (LL) situation;
(2) the lack of tacit knowledge of practising harmony did not stop her from envisioning and carrying the project through;
(3) the use of LL may well be a fruitful alternative in the context of a solo voice project.

Joana Lisboa's interest in LL begun in 2013, largely by watching other artists' solo projects. Initially, Lisboa adopted a simple analogue pedalboard (i.e., footswitch) mainly for singing over loops, but after some years of practice, she then switched to a manual digital table-top board with more commands and inputs available. After five years of practice, Lisboa took her LP to the TV show "The Voice Portugal", thus broadcasting globally her approach to LL in a live situation. From that point on, Joana Lisboa has been using the LP solely for personal and pedagogical purposes.

The live looping setup used in many of her performances, including the compositional process for the aforementioned project, has roughly the same technical configuration:

> The pedals are connected, whether acoustic or electronic, to a sound system (direct speaker or PA) with an XLR cable. Then, I [Joana] connect a Shure M58 microphone to both pedals through another XLR cable. Then, after turning on the pedal, I test the volume and use the LL, changing the presets (which are numerous) according to my taste or performance objective.
> *(Lisboa 2021)*

The implementation of the research took place between November 2020 and March 2021 and consisted in two phases: an online survey (via email) and a semi-structured interview via the online platform Zoom. The survey initially contained ten open questions, being divided into five main areas:

(A) personal information
(B) description of the project
(C) compositional process
(D) technological features
(E) musical influences

After the first survey, another set of questions was submitted, also via email, this time containing only five questions. The intention was to extend the discussion on the technical aspects, including themes such as set up, technology, adaptation, etc. The content analysis was based on a categorical analysis, grounded in a qualitative approach. Surprisingly, side results showed that, alongside the technical issues, other themes from the psychological/philosophical domain came about naturally, steering the discussion to a richer, broader sense. What follows are the findings of the research.

Difficulties

Four major issues were reported by Joana Lisboa: composition, familiarity, interactivity, and autonomy. Right from the beginning, Lisboa pointed out the irregularity of the creative activity as a compositional limitation. According to her:

> The creative process was extremely unstable, because there were very creative days when the songs came up very spontaneously and others when it was only after a few hours of experimenting that an idea that was worth exploring came up. Most of the songs that I find interesting came out in the first 20 days (of 33 days). The remaining days had interesting exploratory moments; however, the results were not always the best.
> *(Lisboa 2021)*

As Lisboa declares, the practical result did not always materialise as a completed composition, for the project allowed moments of free experimentation, creative leisure, and development of unfollowed ideas. In fact, the use of the pedalboard represented a fruitful laboratory to her, as she was able to freely experiment with all kinds of unconventional sounds and techniques.

When questioned about the influences for the project, Lisboa attested that she sought inspiration in ordinary things. From chatting with a friend to casual reading, the wide-ranging scope of influence was clearly related to the circumstances of her inner search for themes and ideas. As she reports, "the songs are autobiographical and reflect all my [Joana's] internal universes".

As mentioned above, a great proportion of the survey emphasises the interaction with the technological apparatus. Beginning with the strategies for controlling the layers within the process of looping them, Lisboa reports that

> Initially, it [i.e., the loop pedal] created so many layers [. . .] that the sound distorted and became unreadable. With the experience of using this technology, I came to realise that everything was related to the balance

between the textures used, the rhythm and the various frequencies represented (high, medium and low sounds).

(Lisboa 2021)

Lisboa describes the challenges of the first contact with the LP acknowledging struggles with timing, pacing, and overlapping layers. Especially when it comes to the physical activity of "pressing the pedal down to define the end of the loop", the need to gain control over this practice seems to represent a form of overcoming personal limitations and uncertainties without losing spontaneity and creativity. Hence, the quality of the first loop is the determinant of the success of the whole song. "The challenge was related to the simplification of the layers and the number of these, in order to represent everything I wanted to include in the songs", she says.

The Zoom interview finished with a broad discussion on the use of technology in music and its relationship with the conditions for good creative practices. For Lisboa, technology is not simply a means to express musicality, but the possibility to "overcome oneself" and to create new paths for creativity. Both her compositional and performative activities benefited largely from this 33-day experience, as she managed to acquire the knowledge and independence for incorporating LL into her musical praxis. Through her own words, "without this technology it would have been much more difficult to express what I [Joana] wanted to express musically". However, despite Lisboa's initial intention of presenting the compositions in a live situation, she opted to play it live only when in the company of other musicians, thus saving the technological tools exclusively for the compositional/recording processes. The reason for this decision lies in the fact that, according to her own words, "it is in sharing with other human beings that I [Joana] identify the psychological, emotional and philosophical dimension of my music".

Lastly, Lisboa acknowledges the desire for a better interaction with electronic devices "due to its unlimited potential of possibilities, especially for a singer who has not yet mastered the piano or the guitar". Going forward, she feels much more confident about other musical projects that might involve the use of LL.

Composition

Typically, the songs were made by juxtaposing vocal layers, with the low voice (i.e., the bass line) being the foundation of the whole song. After the bass, one or two counter melodies in the middle register are added, thus forming the background. By adding simple vocal noises to the vocal parts, different textures and ambiences emerge, thus giving the sense of identity Lisboa sought to portray. Indeed, the background is characterised by setting up not only the rhythmic/harmonic pulse but the mood of the songs.

With those voices in loop, the leading voice (always in a high register) simply lies on top of it. According to Lisboa, this process guarantees "the balance between the various frequencies, as in a choir with sopranos, altos, tenors, and basses".

In a second stage, other percussive details and noises are included, thus allowing an even richer textural character, and bringing a ritualistic approach to the narrative. It is important to mention that, besides the duplication of the process of adding percussive elements, the goal is to arrive at an effective way of recording and looping the voices in which it eases and clarifies the spectrum without the need to add too many layers, thus contaminating the harmony. Lisboa alludes to the creative process as a long section of solo improvisation, "in which the various voices of my being were expressed". The exploratory and highly improvisational sections allowed her critical reasoning to interact with free imagination, thus creating a real laboratory for creation. Only a few elements were predetermined. The structure of the songs was to be loose in length, style, theme, and character. In fact, the only two things that were repeated were the preconceptual method of composition and the recording process.

As it could not be otherwise, the lyrics were entirely improvised. This is another interesting feature, as it confirms the spontaneity and originality of the project. Lisboa asserts that the lyrics of the songs "appeared suddenly", though after the groundwork was already in shape. Metaphorically speaking, the absence of previously chosen lyrics seems to be related to the possibility of listening to one's internal voices and expressing them freely, as if they had their own speakers.

Once the composition was completed, the recording process passed through an audio editing programme. After recording the various layers, she then edited and mixed the audios, removing the superfluous parts, thus giving life to that day's composition. As the final stage, the new song was then shared on an online audio distribution platform (Soundcloud) and other social networks (Facebook and Tumblr).

What follows next is a scheme of how the method for creating a song typically unfolds:

TABLE 11.1 States of the compositional process

pre-composition	exploratory stage				composition practice		share	
exploratory phase	bass line/ vocal noises	middle line(s)	optional extra percussion elements	critical evaluation	leading part/ lyrics	performance	recording	upload to Soundcloud

Although the stages are strongly permeated with a speculative methodology, the process itself encompasses a musical praxis based on two pillars:

(1) action and critical reflection, and (2) subjectivity and background. Furthermore, the epistemological development suggests an autoethnographic approach, where the researcher, when exploring whilst in contact with the object, is immersed and is a constitutive part of the research.

Below is, in first person, a description of the compositional features Joana defined as a formula for composing.

> In my songs there is no harmonic structure as is usual in other music. The harmony is implied between the various accompanying vocal layers and the main melody with the lyrics. The harmony always remains the same and what really gives the songs a sense of structure are the small differences in the interpretation of the stanzas with lyrics by the voice. Starting from a minimalist circular structure composed of one, two or three melodic lines, as if it were a chorus, I improvise layer upon layer, developing and expanding the sonority through melodic and rhythmic variations, always using the voice instrument, adding new colours. in order to create a stable base for the improvisation of melodic lines, the loops remain the same from beginning to end, which was always the goal of the respective project: to give wings to my vocal creativity and will to improvise musically.
>
> *(Lisboa 2021)*

Next is an initial description of three songs of *Uma música por dia nem sabes o bem que te fazia* highlighting the lyrics, as described by Lisboa herself.[5]

Escrever ("Writing"): Track #7

> "This song first came about as a lyric [only], while I was having a training on playful strategies for developing writing competence in children in primary school. The lyrics that emerged were
>
> > *Write, write with freedom!*
> > *Listening, listening to our inner stories*
> > *and giving them shape, body and soul*
> > *In the form of letters and conversations*
> > *Writing is travelling, writing is flowing,*
> > *writing is living adventures in the same place.*

Once the lyrics took shape, a bass voice that mimics a bass, often associated with jazz, emerged, and became the skeleton of the song. I then added a middle range voice, with a more open and free phrasing, similar to a saxophone. The main melodic line is the highest, in which the lyrics previously described were used for improvisation, until the melody stabilised in a more concrete idea. In the end, a jazzy theme emerged, which gave rise to one of the themes included in the EP *Rascunho*."

Aprender ("Learning"): Track #4

The song started out with a question: how could I craft a set of male voices that belong to the same African tribe in which there is a wiser male element that is the pillar to other intrinsically related voices? Against this background, a man stands out in the tribe, singing the following lyrics:

> Life is always rolling
> All you have to do is roll the dice
> Life's rollin'
> You're here to learn/live.
>
> *(Lisboa 2013)*

The repetitions of the strophes of *Aprender* are permeated with moments of intuitive improvisation, alluding to a ritual mood. Unlike the previous theme, *Escrever*, which underwent some alterations until the main melody was stabilised, Lisboa's "conceptual tribe" comes to life and the theme *Aprender* emerges naturally, with almost no further modifications for the recording of the EP.

E dança muito bem[6] ("And [she] dances very well"): Track #2

Only two songs of the project include percussion instruments: Track #11 encompasses a Jew's harp and Track #2 incorporates a goblet drum (also called darabukka). Below are the lyrics of the song *E dança muito bem*.

> And the odalisque comes and dances very well
> To the steps of the music
> And the ant comes and dances very well
> To the steps of the music
> And the elephant comes and dances very well
> To the steps of the music

Here, the rhythmic activity and lyrics are infused with an oriental atmosphere. The ethnic percussion gives a special character to this song and a good amount of novelty to the set. Overall, the individual style is distinctive, and the set is well balanced, for the two songs that include percussion instruments are set quite apart from each other. Surprisingly, despite being a solo voice and LL project, all the songs of the set contrast quite a lot with each other.

The one and the many **171**

TABLE 11.2 Synthesis of musical parameters of the selected songs

	Harmonic characteristics	Rhythmic characteristics	Formal characteristics	Technical characteristics
E Dança Muito Bem	• Key: D Major; • Middle register; vocal range: B2 to F#3; • Predominantly stepwise motions; • Main melody is in soprano, which carries all the lyrics; remaining voices enhance the melodic activity.	• Common Time; • Tempo: *Allegro Moderato*, with no fluctuations; • Rhythmically active, with large use of dotted figures and syncopations; • Oriental character.	• Form: ‖:A:‖ Coda (4-bar strophes repeated). • Similar musical shape and character throughout; • Modal oscillation (D Major and B minor); • Darabukka and Bass line work as the rhythmic/harmonic foundation.	The voices are layered upon the succession of the 4-bar structure and fluctuate over the harmonic narrative, yet the foundation (lower lines) is kept unaltered throughout.
Aprender	• Key: E minor; • Low register; vocal range: Ab2 to Db3; • Predominantly stepwise motions.	• Common Time; • Tempo: *Andante* • Rhythm: syllabic, prosodic; • The rhythmical disposition of the phrases is, however, mutable; • African style.	• Form: ‖:A:‖ Coda; • Harmonically driven; • Composed upon a single musical sentence; • Lyrics are succinct; • There is a vocal improvisation between each strophe; • Harmonically driven; • Composed upon a single musical sentence; • Lyrics are succinct; • There is a vocal improvisation between each strophe; • Unusual harmonic progression: Eb minor, Ab major, Bb minor.	The song is based on a harmonic structure created upon three melodical lines with different rhythms and pitches (middle and low register). The loops begin at the lowest line, adding more texture and complexity as the strophes repeat.
Escrever	• Key: D major; • Lowest register; vocal range: G2 to Bb3; • Predominantly stepwise motions; • Main melody is in G minor.	• Common Time; • With swing; • Tempo: *Allegro* • Jazz feeling.	• Form: ‖:A:‖ Coda; • Harmonically driven; • Composed upon a single musical sentence; • Lyrics are succinct; • There is a vocal improvisation between each strophe.	The two melodic lines (low and middle register) recorded at the beginning of the song are kept in loop until the very end of the song.

Next is a table with a synthesis of some standard parameters related to these three songs. In the far-right column can be found basic information about the applications and corollaries of live looping.

Regarding the application of the LP, the layering process is, thereafter, manifestly cumulative, i.e., once the basses, tenors, altos, and percussive noises are in place and running, they are all kept intact throughout, and become the basis for the higher, most active part. This means that the soprano part was never considered as "loopable" material whatsoever; it appears only later as a response to what the loop cycles are proposing. Inspiration for lyrics then comes, in most cases, from listening to and the reaction to the mosaic of compositional materials disposed in loop. This strategy by no means brings down the quality of the other voices. Rather, it seems to be a way to build and set up a strong base for vocal improvisation by using a wider range of frequencies, as well as bringing a sense of a lively, uplifting character to the songs.

Lastly, a note on the vocal noises. Instead of simply imitating drums and other percussion instruments, the goal for this project was to enrich the accompaniment and dialogue with the melodic looped lines. By contrasting with low frequencies, since the noises are normally short and high in pitch, these sounds were designed to increase the polyphonic texture. At the end of 33 days, Lisboa felt that the project had already served its purpose and decided to put an end to it.

Conclusion

"Through the voice pedal [i.e., loop pedal], I managed to give birth to the various voices that inhabited me and to vent my creativity and latent musicality". This single phrase synthetises the main goal of this article, i.e., to fuse the concepts of heteronyms and LL. The use of the voice served to convey and, specifically in this case, to restore the identity the artist manifests in every creation. For Lisboa, the voice was the raw material for expressing herself both musically and psychologically. The challenge of using mainly vocal tones was, therefore, a deliberate attempt to surpass her own fears and limitations.

> Clearly, the various voices in my music are a product of the various characters that cohabit within me. There are the voices of ancestral men, the voices of children who want to explore the unknown, the voices that want to be other instruments, among many others. Above all, I feel that they are a group of beings that come together at that moment and co-create something beautiful and meaningful.
>
> *(Lisboa 2021)*

The inclusion of the heteronyms in the discussion was an attempt to link the pluralism of the genius of Fernando Pessoa into a world of equal complexity and personal confrontations. LL allows the possibility to confront oneself and to deal with multiple forms of artistic creations displayed side by side.

It is important to bear in mind that the project was based entirely on vocal experimentation. On this account, improvisation became not only a tool with which to reveal Lisboa's musical language, but also the very media in which she could analyse herself in the search for her identity. The past becomes a fundamental aspect of her expressiveness. In the pursuit of her own voice, Lisboa found that she was not alone; she had the background of many years of practice, the voices of many singers and composers she had come across in life. Although it may have felt uncomfortable at first, this project was the decisive test for discovering independence and the sense of freedom.

Ultimately, the relationship between the songwriter Joana and the person Joana is complete, as there was no one else involved. However, as Joana herself puts it, "as with all teamwork, there is not always a balance". The balance was expressed in the search for the combinations that made the most sense in the moment of creation.

> All voices come out of my heart and from the desire to find the balance already mentioned. This project helped me to access more hidden and deeper parts of my being and for that I am extremely grateful.
>
> *(Lisboa 2021)*

Lisboa's songs are clearly an extension of herself. "I [Joana] changed the way I saw myself and understood the creative wealth that exists within me, which is infinite in some way". Nowadays, Joana is a vocal coach in Lisbon, Portugal, where she teaches students to sing and to use their spoken voice.

Notes

1 To understand the concept of heteronyms applied to Pessoa's works, the reader may check Richard Zenith's *Pessoa: Uma biografia* (2022), a 1184-pages-long biography recently published by Quetzal Eds.
2 Seventy-five, to be precise (c.f. MAUNSELL, 2012).
3 www.soundcloud.com/joanalisboa.
4 From here on, all quotes are from Joana Lisboa.
5 The complete list of the songs, including titles, lyrics, and short descriptions can be found at https://www.tumblr.com/blog/view/umamusicapordiaobemquetefazia.
6 A live performance of this song can be found at https://www.youtube.com/watch?v=2qmkEkaDaPw.

References

Lisboa, Joana. "Homepage of the project". Summer 2013. https://www.tumblr.com/blog/view/umamusicapordiaobemquetefazia.
Lisboa, Joana. "Online survey 1". November 2020.
Lisboa, Joana. "Online survey 2". June 2021.
Maunsell, Jerome Boyd. "The hauntings of Fernando Pessoa". *Modernism/Modernity* 19, no. 1 (2012): 115–137.
Peruzzolo-Vieira, Samuel. "Zoom interview with Joana Lisboa". March 17, 2021.
Pessoa, Fernando. *Correspondência (1923–1935)*. Manuela Parreira da Silva (ed.) Lisboa: Assírio & Alvim, 1999), 337–348.

INDEX

Locators in **bold** refer to tables and those in *italics* to figures.

3 AM Sessions (Oliveira) 133, 134

Ableton Live (AL) 87, 88
"affordances" 35, 95, 97, 100, 105, 153
Afro-Jazz 146–147
agency, using the loop pedal 157–159
Akiré 62–63
Almeida, Hernani 146
Almeida, Jorge 137–147
Alperson, Philip 15–16, 27–28
amplification 19, 55, 74, 102
"And [she] dances very well" (Lisboa) 170
Andersson, Teresa 131
Ângelo, Tiago 46
Armorial music 3, 6, 10, 67–71, 75–77
audience: Jorge Almeida's relationship with 145–146; studio and live performances 24–28; weight of a sound 107–109
augmentation 35, 39, 41; *see also* Hybrid Augmented Saxophone of Gestural Symbiosis (HASGS)
Auslander, Philip 26–27

Barney, Katelyn 154, 160
Barth, Michael Edwin 93
bass drum 18, 140
Bennett, Joe 152–153, 160
berimbau 68
Bittencourt, Luís 13–30

Bluetooth connectivity, HASGS 37–38
body capability 141
body posture 74
Bona, Richard 53–54
Brazilian musical traditions 66–68
Burgess, Joby 16, 17–18, 26
Burnard, Pamela 149, 154, 160

Cabral, Ricardo 137–147
Calvino, Italo 111
Canot, Nicolas 43
Cape Verdean music 137–139, 146–147
Castanheira, José 127–135
Cenário do Porto (concert) 60–63
Chapman, Dale 130–131, 132–133
Cherry, Amy K. 93
chord progressions 111–112, 136
"Cicadas Memories", HASGS 44–46
collaborative texts 7–9
community of practice 2; *see also* loopers
competence *see* performance skills
composition 4–5; composer-performer collaboration for *Quasitude* (2020) 78, 81–82; as creative process 106, 114–119, 127–128; *Densus Bridge* 94–95; gestures and structures 107, 112–114, 115–116; impact of live looping 6, 52–65; improvisation 107, 110–111; irregularity of the creative process 166; live looping

competitions 109–110; stages of process 52–65, 167–172, **168**; tonal orientation 131–132; using the loop pedal 6, 109, 111–112, 149–155; *viola sertaneja* 68–69
"Comprovisação nº9", HASGS 47–48
Cook, Nicholas 149, 152, 153
COVID-19 pandemic 51–52, 61, 156
creativity: irregularity of the creative process 166; of Jorge Almeida 139–143; loopers 52–65; "Serpente Infinita" (Valente) 114–119; songwriting 152–153; of Tiago Oliveira 127–135; using the loop pedal 159–160
cyborgization 3–4

de Sousa, Iury 51–65
Densus Bridge 91, 94–104
Digital Audio Workstation (DAW) 87, 154–155
digital devices, in live performance 24–28
digital music instruments (DMIs) 13, 25–28, 34–35
"Disconnect", HASGS 42–44
distributed subjectivity 131–132
Dludlu, Jimmy 146
dodecaphonism 90
drones in music 96–97
Duarte, Alexsander 1–9, 66, 90–104

"Eduardo" (Valente) 114–119
electric guitar: loop features 54–59; use of loops 66
Electronic Wind Instrument (EWI) 35–36; *see also* Hybrid Augmented Saxophone of Gestural Symbiosis (HASGS)
Eno, Brian 1

Fripp, Robert 1
"frippertronics" 1

Garage-Band (GB) 87, 88
gestural performance 129–130
gesture-sound causality 27
gestures 107, 112–114, 116
Gomes, Elielson 90–104
the guitar: as an instrument 142; loop features 54–59; relationship with the loop pedal 138, 142–143

Haraway, Donna 4
Hardjowirogo, Sarah-Indriyati 14–15, 26–27, 29, 83

HASGS *see* Hybrid Augmented Saxophone of Gestural Symbiosis (HASGS)
heteronyms, in Pessoa's work 163–164, 173
Hewitt, Donna 3, 27, 159
Hiney, Aoife 149–160
Hoad, Catherine 150–151, 158
Hossn, Munir 54
Hulme, Pamela 150–151
Hybrid Augmented Saxophone of Gestural Symbiosis (HASGS) 5, 35–36; evolution 36–38, 48–49; instrumental technique 39; mapping 38–39, *41*, **43**; repertoire 39–48

Impett, Jonathan 152
Import/Export: Percussion Suite for Global Junk 16–30; instrumental potential of looper devices 13–14; performer's perspective 18–19; from studio to stage 24–28; use of live looping 19–22; whys and whens of live looping 22–24
improvisation: jazz music 80, 134–135; as research 107, 110–111, 116; spontaneity and planned sections 144–145; in Tiago Oliveira's music 132–134
"Indeciduous", HASGS 40–42
instrumentality: digital music instruments (DMIs) 13, 25–28, 34–35; exploration and expansion of the instrument 110–111; reflections on 5, 14–15, 28–29; sound exploration on the xylophone 82–84
"Invasão" (Valente) 112–113

jazz music: Afro-Jazz 146–147; improvisation 80, 134–135; as influence 52, 169–170; rehearsal approach 155
Junior, Votta 96
junk objects, used in performance 16; *see also Import/Export: Percussion Suite for Global Junk*; instrumentality

Kaschub, Michele 158
Knowles, Julian 3, 27, 159
Kosmik (Oliveira) 127, 128

"Learning" (Lisboa) 170
Lisboa, Joana 164–173
listeners *see* audience
live creation as process 129–130
live looping competitions 109–110
live looping (LL): and the artist 51–52; *Import/Export: Percussion Suite for Global Junk* 16–22; technologies 2–3; use of term 1–2
live sound processing (LSP) 16–18
"liveness" 26–27
loop artists 2; *see also* loopers
loop extension 140–141
loop pedal 3; agency 157–159; composition using 6, 109, 111–112, 149–155; creativity with 159–160; *Densus Bridge* 99–101; flexibility in use of 102–103; and the guitar 138, 142–143; *Import/Export: Percussion Suite for Global Junk* 20, 26; in Jorge Almeida's music 138, 142–143, 146; live looping and the artist 53; *Oração* (Novella) 156–157; "Repente" 72–73, 74; in Tiago Oliveira's music 127–128, 134–135
loopers: blending of humans with machines 3–4; as community of practice 2; creative process 52–65; HASGS 40–41; live looping and the artist 51–52; *see also* performance skills
LoopLab 2–3, 66, 80, 91
Loueke, Lionel 54
Louzeiro, Pedro 47

Madureira, Antonio 69–71
Maier, Florian Magnus 94
Marchini, Marco 131, 134–135
the marimba 68, 72
Marrington, Mark 154–155
McNutt, Elizabeth 101–102, 103
Mendes, Helvio 78–88
microphones: "on-off" switch 60; pedal microphone 101
Mingus, Charles 155
minimalism 1, 96
modal music scales 69, 69–70
Museu de Arte, Arquitectura e Tecnologia (MAAT) 120
musical sensitivity 130–131
Myo Armband 36–37, 38

Nicolls, Sarah 129–130
Nilsson, Per A. 36
NIME community 35–36
north-eastern mode 69–70
notation: in different genres 152–153; HASGS 39–40, 41, 42, 46, 46; proportional 84, 86; "Repente" 71–72, 74
Novella, Isabel 149–160

Oliveira, Tiago 127–135
"on-off" switch 60
one-man bands 2; *see also* loopers
Oração (Novella) 156–157
ostinati 17, 20
overdubs 59–60

pedal chain 56, 64
pedal microphone 101
performance recordivity 3
performance skills: composer-performer collaboration 78, 81–82, 155; *Densus Bridge* 101–102; HASGS 40–41, 43–44; *Import/Export: Percussion Suite for Global Junk* 20–22, 29–30; improvisation 107, 110–111, 116, 132–134; live looping and the artist 51–52, 129–132; loop pedal 144–145; training and practice 141–142
Pessoa, Fernando 163–164, 173
polyphony, *Import/Export: Percussion Suite for Global Junk* 23–24
Portovedo, Henrique 34–49
posture 74
practice time 141–142
programming language, HASGS 40
Prokofiev, Gabriel 16–19; *see also Import/Export: Percussion Suite for Global Junk*
Pronk, Erik 3, 66–76
proportional notation 84, 86

Quasitude (2020) 78–88; acoustic and electronic media 85–88; composer-performer collaboration 78, 81–82; notation 84, 86; sound exploration on the xylophone 82–84
Quinteto Armorial 68–69, 71, 75

real-time performance 3
the rebec 68
REC Pedal Activation (RECORD) 75
rehearsal time 141–142

religious influence, *Oração* (Novella) 156–157
repetition 72; live looping and the artist 63; "Repente" 71–72; in Tiago Oliveira's music 133–134
Riley, Terry 1, 90
robotization 3–4
Rui, Laura 54

Sardo, Susana 1–9
Schafer, Raymond Murray 1
Schoenberg, Arnold 90
Sci-Fi/Ambient Improvisation (Oliveira) 134
"Serpente Infinita" (Valente) 114–119
Sheeran, Ed 94
Sheppard, Philip 151
Siegel, Steven 93
signal quality 57
Silva, Melina 127–135
skill of performer *see* performance skills
Smith, Janice 158
sound atmospheres 128–135
"sound producing devices" 14–15
"soundscapes" 1, 51
space-time continuum 107
spontaneity 144–145; *see also* improvisation
Stasi, Carlos 83, 85, 88
structures, in composition 107, 112–114, 115–116
Suassuna, Ariano 67, 68

Tavares, Gonçalo M. 120
technologies: music industry 2–3; reflections on instrumentality 14–15; studio and live performances 24–28
tempo 143–144
Théberge, Paul 155, 160
threshold technologies 25
"Time-Lag Accumulator" system 1, 3
Tobias, Evan 153–154, 160
tonal orientation 131–132; *see also* modal music scales
Toynbee, Jason 155
trumpet: *Densus Bridge* 94–104; use of electronic devices with 91–93
Turino, Thomas 3
twelve-tone serialism 90

Valente, José 106–120
"Verisimilitude", HASGS 46–47
Vieira, Samuel 78–88, 163–173
the viola 66–67, 71–72, 76
viola sertaneja 66–76; arrangement of "Repente" 69–76; compositional style 68–69
virtuosity 130–131

weight of a sound 107–109
Western art music 16–17, 24, 29–30, 152
Wi-Fi connectivity, HASGS 37–38
Wilson, Oli 150–151, 158
"Writing" (Lisboa) 169–170

XyLoops 80
xylophone 78–88

Yang, Caleb 151
YouTube 51, 54, 62–63, 94, 133–135